I hope this book will serve as an introduction to some of the many diversified facets of mathematics, and that the reader will be encouraged to explore in greater depth the ideas presented.

— Theoni Pappas

THE JOY OF MATHEMATICS

Discovering Mathematics
All Around You

by theoni pappas

**nature • science • architecture • art
history • literature • philosophy • music**

Wide World Publishing/Tetra

Revised Edition 1989

1st   Printing 1986
2nd Printing 1987
3rd Printing May 1989
4th Printing September 1989
5th Printing March 1990

Wide World Publishing/Tetra
P.O. Box 476
San Carlos, CA 94070

Library of Congress Cataloging–in–Publication Data

Pappas, Theoni.
        The Joy of Mathematics / by Theoni Pappas. --Rev. ed.
              p.   cm.
        Includes index.
        ISBN: 0-933174-65-9 : $10.95
        1. Mathematics -- Popular works.   I. Title.
QA93. P365   1989
510 -- dc20   89–16503
                                                                       CIP

"...the universe stands continually open to our gaze, but it cannot be understood unless one first learns to comprehend the language and interpret the characters in which it is written. It is written in the language of mathematics, and its characters are triangles, circles, and other geometric figures, without which it is humanly impossible to understand a single word of it; without these, one is wandering about in a dark labyrinth."
—Galileo

The Joy Of Mathematics unveils concepts, ideas, questions, history, problems, and pastimes which reveal the influence and nature of mathematics.

To experience the joy of mathematics is to realize mathematics is *not* some isolated subject that has little relationship to the things around us other than to frustrate us with unbalanced check books and complicated computations. Few grasp the true nature of mathematics — so entwined in our environment and in our lives. So many things around us can be described by mathematics. Mathematical concepts are even inherent in the structure of living cells.

This book seeks to help you become aware of the inseparable relationship of mathematics and the world by presenting glimpses and images of mathematics in the many facets of our lives.

The joy of mathematics is similar to the experience of discovering something for the first time. It is an almost child-like feeling of wonder. Once you have experienced it you will not forget that feeling — it can be as exciting as looking into a microscope for the first time and seeing things that have always been around you that you have been unable to see before.

When deciding how to organize *The Joy Of Mathematics*, at first certain divisions immediately came to mind—for example, mathematics and nature, mathematics and science, mathematics and art, and so forth. But mathematics and its relationship to our surroundings does not come already packaged into categories. Rather, mathematics and its occurrences are spontaneous with elements of surprise. Thus, the topics are randomly arranged to retain the true essence of discovery. *The Joy Of Mathematics* is designed to be opened at any point. Each section, regardless of how large or small, is essentially self-contained.

After one experiences the sheer joy of mathematics, the appreciation of mathematics follows and then the desire to learn more.

"There is no branch of mathematics, however abstract, which may not someday be applied to the phenomena of the real world."

—Lobachevsky

TABLE OF CONTENTS

TABLE OF CONTENTS

TABLE OF CONTENTS

Mathematics is a science, a language, an art, a way of thinking. Appearing in nature, art, music, architecture, history, science, literature — its influence is present in every facet of the universe. *Mathematics* is a science, a language, an art, a way of thinking. Appearing in nature, art, music, architecture, history, science, literature — its influence is present in every facet of the universe. *Mathematics* is a science, a language, an art, a way of thinking. Appearing in nature, art, music, architecture, history, science, literature — its influence is present in every facet of the universe. *Mathematics* is a science, a language, an art, a way of thinking. Appearing in nature, art, music, architecture, history, science, literature — its influence is present in every facet of the universe. *Mathematics* is a science, a language, an art, a way of thinking. Appearing in

The Evolution of BaseTen

Early forms of counting had no positional base value system[1]. But around 1700 B.C. the positional base 60 evolved. It was very helpful to the Mesopotamians who developed it to use in conjunction with their 360 day calendar. The oldest known true place value system is that devised by the Babylonians, and was derived from the Sumerian sexigesimal system. Instead of needing sixty symbols to write the numerals from 0 to 59, two symbols — Y for 1 and ⟨ for 10 — sufficed. Sophisticated mathematical computations could be performed with it, but no symbol for zero had been devised. To

$$- = \equiv \curlyvee \mathsf{r} \, \mathit{6} \, 7 \, \mathit{5} \, \mathit{7}$$

HINDU (Brahmi)– c. 300 B.C.

$$\mathit{7} \, \mathit{?} \, \mathit{2} \, \mathit{8} \, \mathit{4} \, \mathsf{L} \, \mathit{7} \, \mathsf{r} \, \mathit{9} \, \circ$$

HINDU (Gwalior)– 876 A.D.

$$\mathit{8} \, \mathit{7} \, \mathit{3} \, \mathit{8} \, \mathit{y} \, \mathit{\xi} \, \mathit{v} \, \mathit{7} \, \mathit{C} \, \circ$$

HINDU (Devanagari)– 11th century

$$\mathit{1} \, \mathit{2} \, \mathit{3} \, \mathit{4} \, \mathit{5} \, \mathit{6} \, \mathit{7} \, \mathit{8} \, \mathit{9}$$

WEST ARABIC (Ghobar)– 11th century

$$\mathit{/} \, \mathit{r} \, \mathit{r} \, \mathit{r} \, \mathit{o} \, \mathit{y} \, \mathit{V} \, \mathit{\Lambda} \, \mathit{9} \, \cdot$$

EAST ARABIC– 1575

$$\mathit{1} \, \mathit{2} \, \mathit{3} \, \mathit{2} \, \mathit{4} \, \mathit{5} \, \mathit{\wedge} \, \mathit{8} \, \mathit{9} \, \circ$$

EUROPEAN– 15th century

$$\mathit{1} \, \mathit{2} \, \mathit{3} \, \mathit{4} \, \mathit{5} \, \mathit{6} \, \mathit{7} \, \mathit{8} \, \mathit{9} \, \bullet$$

EUROPEAN– 16th century

$$\mathsf{1} \, \mathsf{2} \, \mathsf{3} \, \mathsf{4} \, \mathsf{5} \, \mathsf{6} \, \mathsf{7} \, \mathsf{8} \, \mathsf{9} \, \mathsf{0}$$

COMPUTER NUMERALS– 20th century

indicate zero an empty position was left in the number. About 300 B.C. a symbol for zero appeared, ⨉ or ⋏ , and the base 60 system developed extensively. In the early A.D. years, the Greeks and Hindus began to use base 10 systems, but they did not have positional notation. They used the first ten letters of their alphabet for counting. Then around 500 A.D. a Hindu invented a positional notation for the base 10 system. He abandoned the letters which had been used for numerals past 9, and standardized the first nine symbols. About 825 A.D., the Arab mathematician *Al-Khowavizmi* wrote an enthusiastic book about the Hindu numerals. The base 10 system reached Spain around the 11th century when the Ghobar numerals were formed. Europe was skeptical and slow to change. Scholars and scientists were reticent to employ the base 10 system because it had no simple way to denote fractions. But it became popular when merchants adopted it, since it proved so invaluable in their work and record keeping. Later, decimal fractions made their appearance in the 16th century, and the decimal point was introduced in 1617 by John Napier.

Someday, as our needs and ways of computing change, will a new system evolve and replace base ten?

---

[1]A positional base value system is a number system in which the location of each digit influences the value of that digit. For example, in base ten for the number 375, the 3 digit is not merely worth 3 but because it is in the hundreds place it is worth 300.

The Pythagorean Theorem

Anyone who has studied algebra or geometry has heard of the *Pythagorean Theorem*. This famous theorem is used in many branches of mathematics and in construction, architecture and measurement. In ancient times, the Egyptians used their knowledge of this theorem to construct right angles. They knotted ropes with units of 3, 4 and 5 knot spaces. Then, using the three ropes, they stretched them and formed a triangle. They knew the triangle would always end up having a right angle opposite the longest side $(3^2+4^2=5^2)$.

Pythagorean theorem :
Given a right triangle, then the square of the hypotenuse of a right triangle equals the sum of the squares of the two legs of the right triangle.

$$a^2 + b^2 = c^2$$

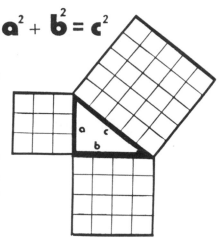

Its converse is also true.
If the sum of the squares of two sides of a triangle equals the square of the third side, then the triangle is a right triangle.

Although this theorem is named after the Greek mathematician, Pythagoras (circa 540 B.C.), evidence of the theorem goes back to the Babylonians of Hammurabi's time, over 1000 years before Pythagoras. Perhaps the name is attributed to Pythagoras because the first record of written proofs come from his school. The existence of the *Pythagorean theorem* and its proofs appear throughout continents, cultures and centuries. In fact more proofs of this theorem have probably been devised than any other!

G raphics is another area in which people are exploring the use of computer. The optical illusion below is a computer rendition of *Schroder's staircase*. It falls in the category of an oscillating illusion. Our minds are influenced by past experiences and by suggestions. The mind at first will see an object one way, and when a certain length

Optical Illusions and Computer Graphics

of time has passed it will change its point of view. The time factor is influenced by our attention, or how quickly we get bored with what we initially focus on. In Schroder's illusion the staircase will appear to flip upside down.

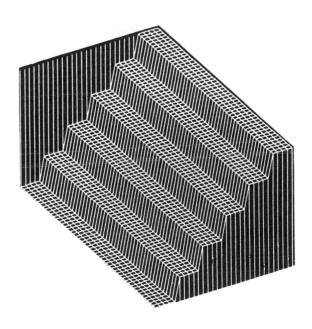

The Cycloid – The Helen of Geometry

The cycloid is one of the many fascinating curves of mathematics. It is defined as —

the curve traced by the path of a fixed point on a circle which rolls smoothly on a straight line.

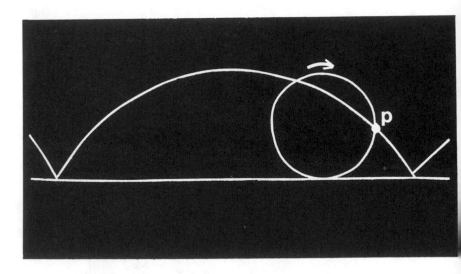

One of the first references to the cycloid appears in a book by Charles Bouvelles published in 1501. But it was in the 17th century that a large number of prominent mathematicians (Galileo, Pascal, Torricelli, Descartes, Fermat, Wren, Wallis, Huygens, Johann Bernoulli, Leibniz, Newton) were intent on discovering its properties. The 17th century was a time of interest in the mathematics of mechanics and motion, which may explain the keen interest in the cycloid. Along with the many discoveries at this period of time, there were many arguments as to who discovered what first, accusations of plagiarism, and minimization of one another's work. As a result the cycloid has been labeled *the apple of discord* and *the Helen of geometry.* Some of the properties of the cycloid discovered during the 17th century are:

1) *Its length is four times the diameter of the rotating circle. It is especially interesting to find that its length is a rational number independent of* π.

2) *The area under the arch is three times the area of the rotating circle.*

3) *The point on the circle that is tracing the cycloid takes on different speeds —in fact at one place, P_5, it is even at rest.*

4) *When marbles are released from different points of a cycloid shaped container, they arrive at the bottom simultaneously.*

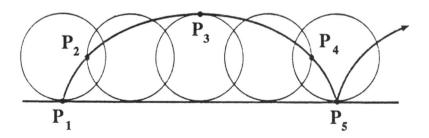

Each circle represents a quarter turn of the rotating circle. Note the length of the quarter turn from P_1 to P_2 is much shorter than from P_2 to P_3. Consequently, the point must speed up from P_2 to P_3 since it must travel further in the same length of time. The point is at rest at where it must change directions.

There are many fascinating paradoxes tied to the cycloid. The *train paradox* is especially intriguing.

> *At any instant a moving train never moves entirely in the direction the engine is pulling it. There is always part of the train moving in the opposite direction than that in which the train is moving.*

This paradox can be explained by using the cycloid. Here the curve formed is called a *curtate cycloid* – the curve traced by a fixed point outside a revolving wheel. The diagram shows that part of the train wheel moves backwards as the train moves forward.

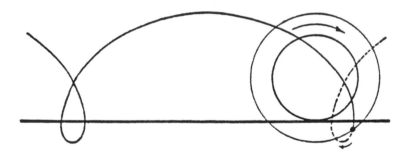

German mathematician, David Hilbert (1862-1943), first proved that any polygon can be transformed into any other polygon of equal area by cutting it into a finite number of pieces.

A triangle to a square

This theorem is illustrated by one of the puzzles of the renowned English puzzlist, Henry Ernest Dudeney (1847-1930). Dudeney transforms an equilateral triangle into a square by cutting it into four pieces.

Here are the four pieces. Fit them back together. First make them into an equilateral triangle and then into a square.

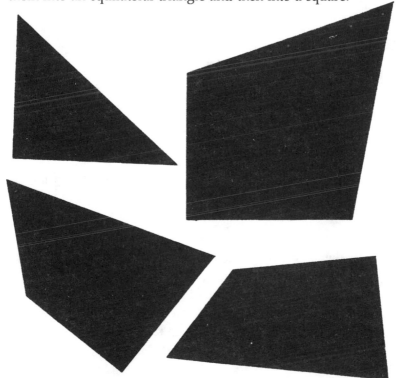

See appendix for solution of *a triangle to a square.*

Halley's Comet

Orbits and paths are concepts that can be easily described mathematically by equations and their graphs. Studying graphs can sometimes reveal cycles and periods of paths. And so it must have been in the case of *Halley's Comet.*

Halley's comet as depicted in the Bayeaux tapestry

Until the 16th century, comets were an unexplained astronomical phenomena which did not seem to obey the solar system laws of Copernicus and Kepler. But in 1704 Edmund Halley worked on the orbits of various comets for which data was available. Some of the most comprehensive records were on the comet of 1682. He noticed that its orbit passed through the same regions of the sky as the comets of 1607, 1531, 1456, and concluded they were a single comet orbiting the sun elliptically about every 75 to 76 years. He successfully predicted its return in 1758, and it became known as **Halley's Comet**. Recent research suggests that Halley's Comet may have been recorded by the Chinese as early as 240 B.C.

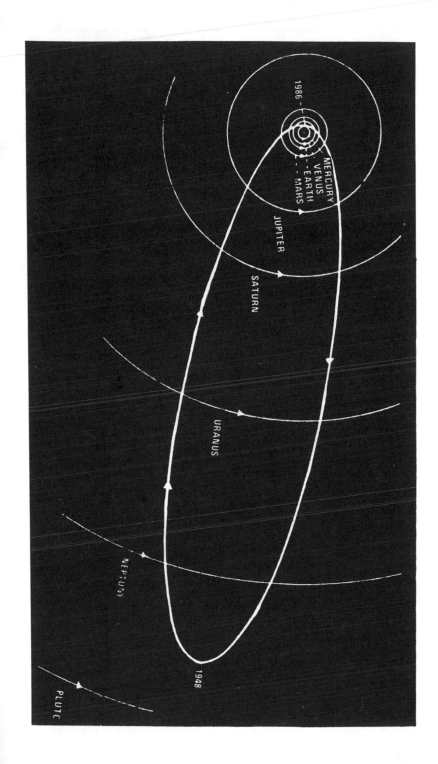

With each appearance, the spectacle of Halley's tail fades as evidenced by its last appearance in 1985-1986.

Comets are believed to originate from ice planetoids which circle the sun in a spherical shell about 1 to 2 light years from the sun. These planetoids are composed of ice and particles of metal and silicates. Here at the outskirts of our solar system, at freezing temperatures, these planetoids circle the sun at speeds of 3 miles per minute, thus taking 30,000,000 years to orbit the sun. Occasionally gravitational interference from nearby stars slows a planetoid causing it to fall toward the sun, thereby changing its circular orbit to an elliptical one. Once it begins its elliptical orbit about the sun, part of its ice becomes gaseous. This forms the comet's tail which always points away from the sun because the tail is blown by the sun's solar winds. The comet's tail is composed of gases and small particles that are illuminated by the sun. The comet would continually pass in an unchanging elliptical orbit about the sun if it were not for the gravitational influence of Jupiter and Saturn. Each orbit brings the comet closer to the sun, which melts more ice and extends the tail. The tail makes the size of the comet appear much larger (a typical comet is about 10 km in diameter). In the tail travel meteors, originally embedded in icy layers of the comet. The meteors are remnants of the comet which remain in orbit after the comet has disintegrated, creating a meteor shower when its orbit coincides with the Earth's orbit.

M any designs and illustrations have become familiar and often taken for granted. In the *British Journal of Psychology*, February, 1958, Roger Penrose published his impossible tribar.

The Impossible Tribar

He called it a three-dimensional rectangular structure. The three right angles appear normal, but are spatially impossible.

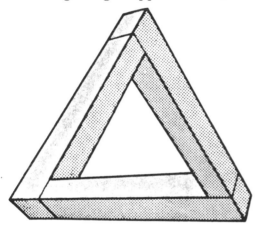

These three right angles seem to form a triangle, but a triangle is a plane object (not three-dimensional) and its angles total 180 degrees, not 270 degrees.

More recently, Penrose has been responsible for the theory of twistors. Although twistors are not visible, Penrose believes that space and time are interwoven by the interaction of twistors.

twister

Can you determine why *Hyzer's optical illusion* is also mathematically impossible?

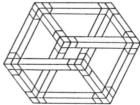

Hyzer's optical illusion

The Quipu

The Inca empire encompassed the area around Cuzco, most of the rest of Peru and parts of Ecuador and Chile. Although the Inca did not have a written mathematical notation system or a written language, they managed their empire (more than two-thousand miles long) by use of **quipus**. **Quipus** were knotted ropes using a positional decimal system. A knot in a row farthest from the main strand represented one, next farthest ten, etc. The absence of knots on a cord implied zero. The size, color and configuration of knots recorded information about crop yields, taxes,

This illustration of a Peruvian quipu was drawn by a Peruvian Indian, D. Felipe Poma de Ayala, between 1583 and 1613. In the lower left hand corner, there is an abacus counting device used with maize kernels on which computations were performed and later transferred to the quipu.

population, and other data. For example, a yellow strand might represent gold or maize; or on a population quipu the first set of strands represented men, the second set women, and the third set children. Weapons such as spears, arrows, or bows were similarly designated.

Accounting for the entire Inca empire was done by a class of quipu scribes who passed on the techniques to their sons. There were scribes at each administrative level who specialized in an individual category.

In the absence of written records the quipus served as a means of recording history — these historical quipus were recorded by armantus (wise men) and passed on to the next generation, which used them as reminders of stories they had been told.

And thus these primitive computers — **quipus** — had knotted in their memory banks the information which tied together the Inca empire.

The Inca Royal Road extended 3,500 miles from Ecuador to Chile. All aspects of goings on in the vast Inca empire were communicated along the Royal Road via chasquis (professional runners), who were assigned a two miles stretch. They were so familiar with their particular increment of road that they could run it at top speed day or night. They would relay information until it reached the desired location. Their service, coupled with the use of quipus, kept the Inca emperor up to date about population changes, equipment, crops, possessions, possible revolts, and any other pertinent data. The information was relayed on a twenty-four hour basis and was very accurate and current.

Calligraphy, Typography & Mathematics

Architecture, engineering, decoration and typography are a few fields in which geometric principles are applied. Albrecht Dürer lived from 1471 to 1528. During his lifetime, he combined his knowledge of geometry and his artistic ability to create many art forms and art methods. He systematized the construction of Roman letters, which was essential for accuracy and consistency of large letters on buildings or tombstones. These illustrations by Dürer show the use of geometric construction for writing Roman letters.

Today, computer scientists have used mathematics to design software to produce quality typography and graphics. An outstanding example is POSTSCRIPT programming language developed by Adobe Systems (of Palo Alto, CA) to work with the laser printer.

The Wheat &the Chessboard

How many grains of wheat are needed if they are placed on a chessboard in the following manner?

One grain in the first square, two in the second, four in the third, eight in the fourth square, and so on doubling the amount with each new square.

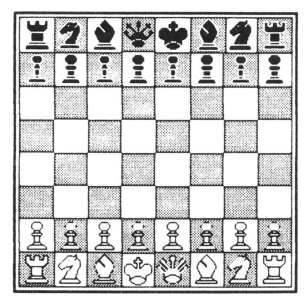

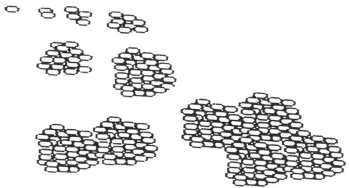

See appendix for solution to *the wheat & the chessboard problem.*

Probability and π

Mathematicians and scientists have always been intrigued by π, but it acquired a whole new following when it foiled a diabolic computer in a *Star Trek* episode. π wears different hats — it is the *ratio* of a circle's circumference to the diameter, it is a *transcendental number* (a number that cannot be the solution of an algebraic equation with integral coefficients).

3.14159265358979323846264 3
3832795028841971693993 75
10582097494459230781640 6
28620899862803482534211 7
06798214808651328230664 7
09384460955058223172535 9
40812848111745028410 27 ...

For thousands of years people have been trying to compute π to more and more decimal places. For example, Archimedes approximated π to between 3 1/7 and 3 10/71 by increasing the number of sides of an inscribed polygon. In the *Bible*, in the *Book of Kings* and in *Chronicles*, π is given as 3. Egyptian mathematicians' approximation was 3.16. And Ptolemy, in 150 A.D., estimated it as 3.1416.

Theoretically Archimedes' method of approximation could be extended indefinitely, but with the invention of calculus, the Greek method was abandoned and the use of convergent infinite series, products, and continued fractions were used to approximate π. For example,

$$\pi = 4/(1+1^2/(2+3^2/(2+5^2/(2+7^2/(\ldots))))).$$

One of the most curious methods for computing is attributed to the 18th century French naturalist, Count Buffon and his *Needle Problem*. A plane surface is ruled by parallel lines, all d units apart. A needle of length less than d is dropped on the ruled surface. If the needle lands on a line, the toss is considered favorable. Buffon's amazing discovery was that the ratio of favorable tosses to unfavorable was an expression involving π. If the needle's length is equal to d units, the probability of a favorable toss is  $2/\pi$. The more tosses, the more closely did the result approximate π. In 1901, the Italian mathematician Lazzerini made 3408 tosses, giving π's value as 3.1415929 – correct to 6 decimal places. In yet another probability method to compute π, R. Charles, in 1904 found the probability of 2 numbers (written at random) being relatively prime to be $6/\pi$.

It's startling to discover the versatility of π, crossing as it does the wide spectrum of geometry, calculus and probability.

Earthquakes and Logarithms

There seems to be a human need to describe natural phenomena in mathematical terms. Perhaps this is because we want to discover methods by which we can have some control — possibly through prediction — over nature. And so it is with earthquakes. It may at first seem unusual to

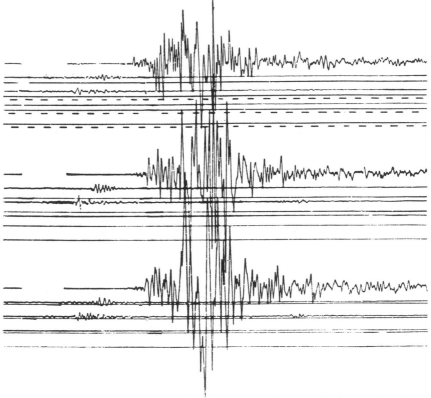

a seismograph of an earthquake

link earthquakes with logarithms, but the method used to measure earthquakes' magnitudes is responsible for the connection. The Richter scale was devised in 1935 by American seismologist Charles F. Richter. The scale measures the earthquake's magnitude by describing the amount of energy released at the focus of the earthquake. The Richter scale is logarithmic so the energy released increases by powers of 10 in relation to the Richter numbers. For example, an earthquake of magnitude 5 releases 10 times the energy than by one of magnitude 4. Thus, an earthquake of magnitude 8 is **not** twice as powerful as that of magnitude 4, but 10^4 or 10,000 times as powerful.

The Richter scale numbers range from 0 to 9, but theoretically there is no upper limit. An earthquake of magnitude greater than 4.5 can cause damage; severe earthquakes have magnitudes greater than 7. For example, the Alaskan earthquake of 1964 was 8.4 on the Richter scale and the San Francisco earthquake of 1906 was 7.8.

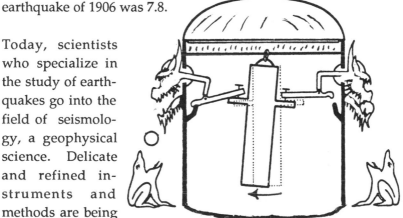

Today, scientists who specialize in the study of earthquakes go into the field of seismology, a geophysical science. Delicate and refined instruments and methods are being sought and devised in order to be able to quantify and predict earthquakes. One of the first instruments and continually used instruments

This representation of the earliest known seismograph was made in China in the 2nd century A.D. from a bronze wine jar six feet in diameter. Eight dragons with bronze balls in their mouths surround the jar. In the event of an earthquake a dragon would drop its ball which would fall in the mouth of a toad below. The instrument would then lock, and thus, indicate the direction of the earthquake.

is the seismograph, which automatically detects, measures and graphically pictures earthquakes and other ground vibrations.

The parabolic ceiling of the Capitol

It seems rather amusing in our present high tech world to find that in the 19th century, the Capitol was coincidently designed with its own non-electronic eavesdropping devices. The United States Capitol was designed in 1792 by Dr. William Thornton, and the structure was reconstructed in 1819 after having been burned in 1814 by invading British troops.

The ceiling in Statuary Hall as it appears today in the United States Capitol.

To the south of the Rotunda (the huge domed hall), is Statuary Hall, so named because in 1864 each state was asked to contribute statues of two of its famous citizens. The House of Representatives met in Statuary Hall until 1857. It was in this room that John Quincy Adams, while a member of the House of Representatives, discovered its acoustical phenomenon. He found that at certain spots one could clearly hear conversations taking place on the other side of the room, while people standing between could hear nothing and their noise did not obscure the sounds coming from across the room. Adam's desk was situated at a focal point of one of the *parabolic reflecting ceilings*. Thus, he could easily eavesdrop on the private conversations of other House members located near another focal point.

Parabolic reflectors function in the following manner:

> *sound bounces off the parabolic reflector (or, in this case a ceiling's dome) and travels parallel to the opposite parabolic reflector, where it bounces to its focal point. Thus all sounds originating at a focal point pass through the opposite focal point.*

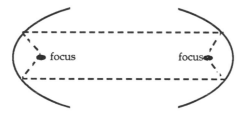

The Exploratorium in San Francisco, California has parabolic sound reflectors set up for public use. They are situated on opposite sides of a large room and their focal points are identified. Two people can carry on a normal conversation speaking at the focal points. Neither the number of people nor the noise level in the room will hinder their ability to hear each other.

Computers, Counting, & Electricity

One communicates with an electronic computer by using a computer language. The computer language in turn is translated into some base system in order to direct the electrical impulses which power the computer. Base ten works very well with our pen and pencil calculations, but another base system is needed for electronic computers. If a memory device were to operate in base ten it would have to assume ten different states for the ten numerals comprising base ten (0, 1,2,3,4,5,6,7,8,9). Although this is possible with a mechanical system, it is not feasible with electricity. On the other hand, the *binary base system* is a perfect candidate for electronic computers. Only two numerals are used in base two (*the binary base*). These are 0 and 1. These numerals can be easily represented by electricity in one of three ways:

(1) by having the current either on or off,
(2) by magnetizing a coil in one direction or the other,
(3) by energizing or not energizing a relay

In any of the three cases, one state is taken by the numeral 0 and the other state by the numeral 1.

Computers do not count the way people do – one, two, three, four, five, six , seven, eight, nine, ten, eleven, twelve,... Instead, they count one, ten, eleven, one hundred, one hundred and one, one hundred and ten, one hundred and eleven,...

Since, computers operate with electricity. Their mechanisms use electricity to translate to symbols we can understand on their monitors. As electricity passes through the intricate parts of a computer, it can either turn a part *on* or *off.* **On and off** are the only **two** possibilities for electricity which is why only **two digits**, 0 and 1, and base two are used by computers.

one, two, three, four, five, six, seven, eight, ...
1, 10, 11, 100, 101, 110, 111, 1000, ...

BASE TEN vs BASE TWO

When we write our numbers we use the digits 0, 1,2, 3, 4, 5, 6, 7, 8, 9. This is called base ten because we use ten digits to form any number. The placement of the digit in the number stands for that digit times a power of ten. When we write our numbers, each digit's value depends on its place in the number, for example,

5374 does not mean 5+ 3+ 7 + 4 , but it means
5 thousands +3 hundreds + 7 tens +4 ones.

Each place in the number is a power of ten:
thousand =1000=10x10x10=10^3
hundred=100=10x10^2
ten=10=10^1
one=1=10^0

Computers write their numbers using only the digits 0 and 1. Their system is called base two because only the first two digits are used to form numbers and each place in the number is a power of two. The first place is the 1's place, then comes the 2's place, then the 2x2=4's place, then the 2x2x2=8's place, and so on.

2x2x2=8's	2x2=4's	2's	1's
2^3	2^2	2^1	2^0

so the number 1101 would mean
1x8 +1x4 +0x2 +1x1 = totals 13 in base ten system.

Topo – A Mathematical Game

Topo is a game of many changing strategies. Any number of people can play. When one is beginning to learn, start with only two players. There are three parts to the game.

> *I.*   *building the regions of play*
> *II.*   *assigning values on some or all regions*
> *III.*   *capturing regions*

I.   Taking turns, each player draws a region adjacent in some way to a previously drawn region. Ten such regions are drawn by each player, as illustrated in **A**.

II.   Each player chooses a colored pen. Then, taking turns, each player assigns a number value to a region until each player colored regions total number value is 100. If a player wants to just assign a region the number 100, then that player would have only one region.

III. *Object of the game:*   The player ending with the most regions is the winner. *Note* the number value of a region is irrelevant.

Moves: A region is captured when one or more adjacent regions belonging to another player total more than that region.

Once a region is captured, it is out of play; and should be marked for the capturing player.

Continue capturing until no more regions can be captured.

Topo has some intriguing variations. The more one plays the more one discovers various strategies in *building regions, assigning values, and capturing regions.*

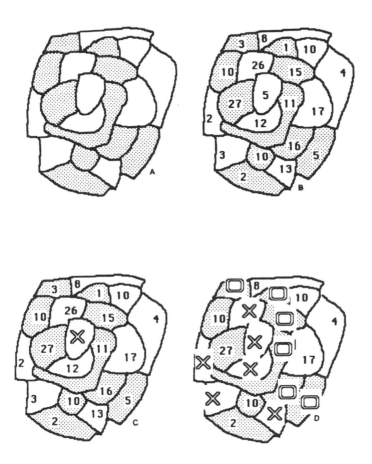

Fibonacci Sequence

Fibonacci[1], one of the leading mathematicians of the Middle Ages, made contributions to arithmetic, algebra and geometry. He was born Leonardo da Pisa (1175-1250), son of an Italian customs official stationed at Burgia in southern Africa. His father's work involved travel to various Eastern and Arabic cities, and it was in these regions that Fibonacci became familiarized with the Hindu-Arabic decimal system, which had place value and used the zero symbol. At this time Roman numerals were still being used for calculating in Italy. Fibonacci saw the value and beauty of the Hindu-Arabic numerals, and was a strong advocate of their use. In 1202 he wrote **Liber Abaci**, a comprehensive handbook explaining how to use the Hindu-Arabic numerals; how addition, subtraction, multiplication and division were performed with these numerals; how to solve problems; and further discussion of algebra and geometry. Italian merchants were reluctant to change their old ways; but through their continual contact with Arabs and the works of Fibonacci and other mathematicians, the Hindu-Arabic system was introduced and slowly accepted in Europe.

Fibonacci sequence — 1, 1, 2, 3, 5, 8, 13, 21, 34, 55, ...

It seems ironic that Fibonacci is famous today because of a sequence of numbers that resulted from one obscure problem in his book, **Liber Abaci**. At the time he wrote the problem it was considered merely a mental exercise. Then, in the 19th century, when the French mathematician Edouard Lucas was editing a four volume work on recreational mathematics, he attached Fibonacci's name to the sequence that was the solution to the problem from **Liber Abaci**. The problem from **Liber Abaci** that generated the Fibonacci sequence is:

---

[1]Fibonacci literally means son of Bonacci.

1) Suppose a one month old pair of rabbits (male and female) are too young to reproduce, but are mature enough to reproduce when they are two months old. Also assume that every month, starting from the second month, they produce a new pair of rabbits (male & female).

2) If each pair of rabbits reproduces in the same way as the above, how many pairs of rabbits will there be at the beginning of each month?

⬤=pair, mature enough to reproduce

◯=pair, too young to reproduce

no. of pairs

1=F1=1st Fib. no.

1=F2=2nd Fib. no.

2=F3=3rd Fib. no

3=F4=4th Fib. no.

5= F5=5th Fib. no.

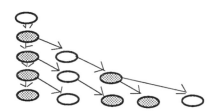

Each term of the Fibonacci sequence is the sum of the two preceeding terms and is represented by the formula:

$$F_n = F_{n-1} + F_{n-2}$$

Fibonacci did not study this resulting sequence at the time, and it was not given any real significance until the 19th century when mathematicians became intrigued with the sequence, its properties, and the areas in which it appears.

Fibonacci sequence appears in:

   I.   _The Pascal triangle, the binomial formula & probability_

  II.   _the golden ratio and the golden rectangle_

 III.   _nature and plants_

 IV.   _intriguing mathematical tricks_

  V.   _mathematical identities_

A twist to the Pythagorean Theorem

Pappus of Alexandria was a Greek mathematician of about 300 B.C. who proved an interesting variation of the Pythagorean Theorem. Instead of dealing with the squares on the legs and the hypotenuse, his theorem dealt with any shaped parallelograms built on the legs and the hypotenuse.

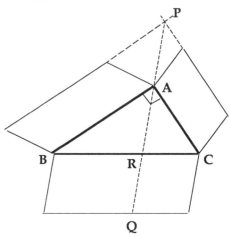

Use any right triangle and the follow procedure:
1) construct any size parallelograms on the two legs of the triangle;
2) extend the parallelograms' sides until they intersect in point P;
3) draw ray PA so that ray PA intersects segment BC at R and |RQ|=|PA|
4) draw a parallelogram on the hypotenuse $\overline{BC}$ so that its two sides are the same length and parallel to $\overline{RQ}$.

Pappus concluded: The area of the parallelogram on the hypotenuse equals the sum of the areas of the other two parallelograms.

What happens if one ring is removed?

Trinity of Rings – a topological model

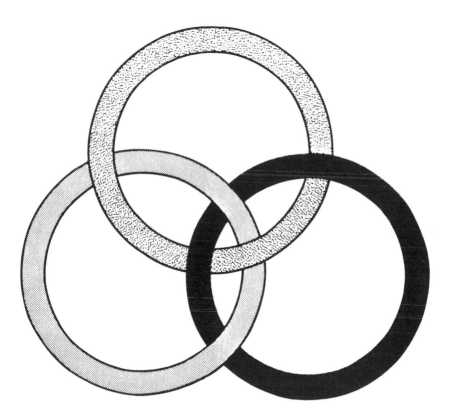

Are any two rings linked?

Are all three rings linked?

Anatomy & the Golden Section

Leonardo da Vinci extensively studied the proportions of the human body. His drawing below has been studied in detail, and shown to illustrate the use of the golden section.[1] This is one of his drawings in the book he illustrated for mathematician Luca Pacioli called *De Divina Proportione* published in 1509.

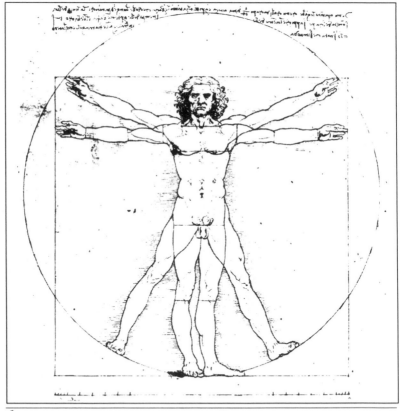

---

[1] The term *golden section* is also referred to as the *golden mean, the golden ratio, the golden proportion* . It is the geometric mean when it is located on a given segment as follows. Point B sections off segment AC so that $(|AC|/|AB|)=(|AB|/|BC|)$. The value of the golden mean may be determined as, $\dfrac{1+\sqrt{5}}{2} \approx 1.6$

A   B   C

The golden section is also present in the unfinished work, *St. Jerome,* by Leonardo da Vinci, painted around 1483. The figure of St. Jerome fits perfectly into a golden rectangle, as superimposed on this drawing. It is believed that this was not an accident, but that Leonardo purposely made the figure conform to the golden section because of his keen interest and use of mathematics in many of his works and ideas. In the words of Leonardo —"*...no human inquiry can be called science unless it pursues its path through mathematical exposition and demonstration.*"

St. Jerome. Leonardo da Vinci. Circa 1483

The catenary & the parabolic curves

A chain hanging freely forms a curve called a catenary[1] curve. The curve looks much like a parabola, and even Galileo first believed it was a parabola.

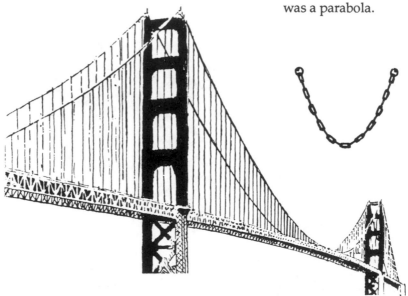

When weights are tied to the catenary curve at equal intervals, the chain becomes a parabolic curve. This is similar to a cable suspension bridge, such as the *Golden Gate Bridge* in San Francisco. Here the parabola is formed when vertical supports are placed on the catenary cable.

The Exploratorium in San Francisco has a hands on exhibit of a catenary arch.

---

[1] The equation of a catenary curve is: $y = a \cosh(x/a)$ where the x-axis is the directrix.

An old but frustrating puzzle is to fit these four pieces together so they form a **T**. Good luck!

The T Problem

See the appendix for the solution to *the T problem*.

Thales & the Great Pyramid

Thales (640-546 B.C.) was known as one of the Seven Wise Men of ancient Greece. Called the *father of deductive reasoning*, he introduced the study of geometry into Greece. He was a mathematician, a teacher, a philosopher, an astronomer, a shrewd businessman and the first geometer to prove his theories by a step-by-step proof. Thales correctly predicted the solar eclipse in 585 B.C., and astounded the Egyptians when he calculated the height of their Great Pyramid using shadows and similar triangles.

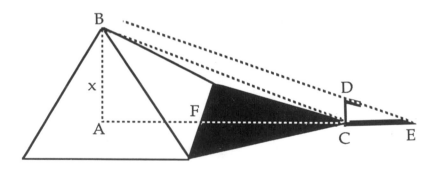

procedure:

The above diagram shows the shadow cast by the pyramid. A rod of known length, |DC|, is placed perpendicularly at the tip of the shadow at C. The rod's shadow has measured length |CE|. |AF| is the length of 1/2 the pyramid's side. Now the height of the pyramid, x, can be easily calculated using similar triangles, ΔABC and ΔCDE.

$(x/|CD|) = (|AC| \, / \, |CE|)$, *thus* $x = (|CD| \cdot |AC|) \, / \, |CE|$

Hotel Infinity

One of the qualifications for hotel clerk at Hotel Infinity is a working knowledge of infinity. Paul applied, was interviewed, and started work the following evening. Paul wondered why the hotel required that all its clerks know about infinity, infinite sets, and transfinite numbers. He figured since it was an infinitely roomed hotel it would be no problem finding rooms for its guests. After his first night on the job, he was glad he had that knowledge.

When Paul relieved the day clerk, she informed Paul that there were an infinite number of rooms presently occupied. As she left, a new guest walked in with a reservation. Paul needed to decide which room to give the guest. He thought for a moment, then moved each occupant to a room with the next highest number, and therefore Paul was able to vacate room #1. Paul felt good about his solution, but just then an infinite bus load of new guests arrived. How would he give them their rooms?

See appendix for *Paul's solution.*

Crystals – Nature's Polyhedra

From classical times polyhedra have appeared in mathematical literature, but their origin is much older, being linked with the origin of nature itself. Crystals grow in the shapes of polyhedra. For example, sodium chlorate crystals appear in the shape of cubes and tetrahedra, while chrome alum crystals are in the form of octahedra. Equally fascinating are the appearance of decahedra and icosahedra crystals in the skeletal structures of radiolaria – microscopic sea animals.

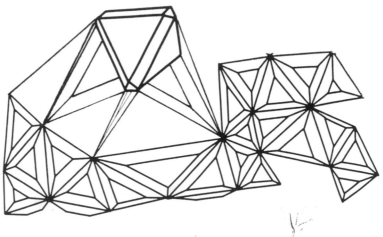

Polyhedra are solids whose faces are in the shape of polygons. They are called *regular polyhedra* if their faces are identical polygons and their angles are identical throughout. Thus, a regular polyhedron has all its faces identical, all its

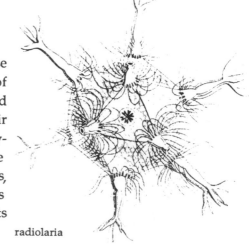

radiolaria

edges identical, and all its corners identical. There are infinitely many types of polyhedra, but there are only five regular
polyhedra — called *Platonic solids*.[1] They are named after
Plato, who discovered them independently about 400 B.C.
Their existence was known before this time by the
Pythagoreans, and Egyptians had used some of them in
architecture and other objects they designed.

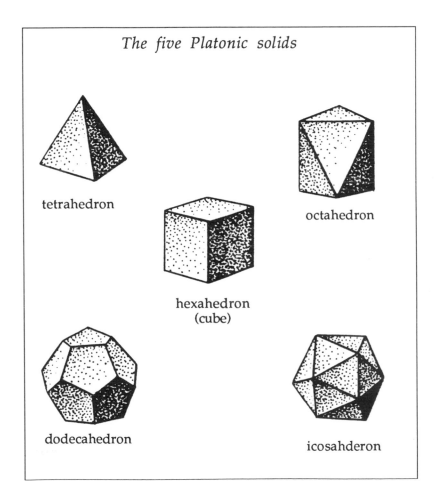

The five Platonic solids

tetrahedron

octahedron

hexahedron
(cube)

dodecahedron

icosahderon

---

[1]See section on **Platonic Solids.**

Pascal's triangle, the Fibonacci sequence & binomial formula

Blaise Pascal (1623-1662) was a famous French mathematician who might have become one of the great mathematicians if it were not for his religious beliefs, poor health and unwillingness to exhaust a mathematical topic. His father, fearing that his son would share his keen interest in mathematics[1] and wanting him to develop a broader educational background, initally discouraged him from studying mathematics in order that he might develop other interests. But by the age of twelve Pascal showed such a gift for geometry that his mathematical inclination was thereafter encouraged. He was very talented, and at the age of sixteen he wrote an essay on conics that surprised and astounded mathematicians. In his work was the theorem that came to be known a Pascal's theorem, which states in essence that *opposite sides of a hexagon, which is inscribed in a conic, intersect in three collinear points.* At the age of eighteen he invented one of the first calculating machines. At this time he suffered from poor health, and made a vow to God that he would give up his work in mathematics. But three years later he wrote his work on the *Pascal triangle* and its properties. On the night of November 23, 1654, Pascal had a religious experience that prompted him to devote his life to theology and abandon mathematics and science. Except for one brief period (in 1658-1659), Pascal never studied mathematics again.

Mathematics has a way of connecting ideas that appear unrelated on the surface. So it is with the Pascal triangle, the

---

[1]Etienne Pascal, was very much interested in mathematics, and in fact the curve *limaçon of Pascal* is named after him rather than his son.

Fibonacci sequence and Newton's binomial formula. The Pascal triangle, the Fibonacci sequence, and the binomial formula are all interrelated. The design illustrates their relationships. The sums of the numbers along the diagonal segments of the Pascal triangle generate the Fibonacci sequence. Each row of the Pascal triangle represents the coefficients of the binomial (a+b) raised to a particular power.

For example,

$(a+b)^0 = 1$ 1

$(a+b)^1 = 1a + 1b$ 1   1

$(a+b)^2 = 1a^2 + 2ab + 1b^2$ 1   2   1

$(a+b)^3 = 1a^3 + 3a^2b + 3ab^2 + 1b^3$ 1   3   3   1

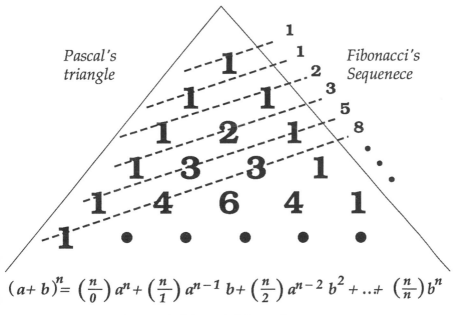

Pascal's triangle

Fibonacci's Sequenece

$(a+b)^n = \left(\frac{n}{0}\right) a^n + \left(\frac{n}{1}\right) a^{n-1} b + \left(\frac{n}{2}\right) a^{n-2} b^2 + ..+ \left(\frac{n}{n}\right) b^n$

Newton's binomial formula

Mathematics of the billiard table

Who would believe that knowledge of mathematics can help one's billiard game? Given a rectangular billiard table whose sides are in whole number ratios, e.g. 7:5, a ball hit from a corner at a 45 degree angle will end up in one of the corners after a certain number of rebounds. In fact, the number of rebounds is tied in with the dimensions of the table. The number of rebounds before reaching a corner is given by the formula:

length + width -2.

STARTING POINT

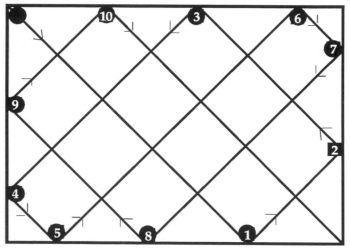

ENDING POCKET
after 10th rebound

With the above table the rebounds total 10.
7+5-2=10 rebounds.

Note the formation of isosceles right triangles in determining the path of the ball.

V arious geometric shapes appear in many facets of the physical world. Many of these shapes are not visible to the naked eye. In this particular electron's path the appearance of pentagons is evident.

The geometry of an electron's path

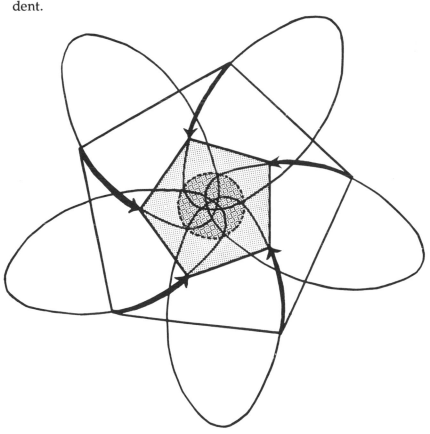

The Moebius Strip & The Klein Bottle

Topologists have created some fascinating objects. The Moebius strip, created by the German mathematician Augustus Moebius (1790-1868), is one such object.

The above diagram illustrates a strip of paper that has been glued to form a band. One side of the paper is on the inside of the band while the other side lies on the outside of the band. If a spider were to crawl along the outside of the band, the only way it could get to the inside section would be to somehow cross its edge.

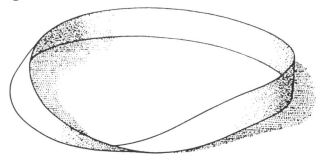

This diagram illustrates the *Moebius strip,* formed by taking a strip of paper and twisting it once before gluing its ends together. Now the paper strip in this form no longer has two sides. It is one sided. Suppose the spider were to begin crawling along the Moebius strip, it would be able to crawl over the entire strip without ever having to cross an edge. To prove this, take a pen and continually draw a line. You will have covered the entire strip and returned to your starting point without lifting your pen.

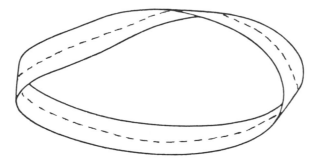

Another interesting property of the Moebius strip is to discover what happens when it is cut along the center mark. Try it.

Moebius bands, as in fan belts for automobiles or belts for mechanical devises, are of particular interest in industry, since they wear more uniformly than conventional belts.

Just as intriguing as the Moebius strip is the Klein bottle. Felix Klein, a German mathematician (1849-1925), devised a topological model of a special bottle which has only one surface. The Klein bottle has an outside but no inside. It passes through itself. If water were poured into it, the water would just come out of the same hole.

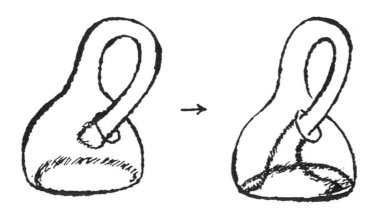

There is an interesting connection between the Moebius strip and the Klein bottle. If the Klein bottle were cut halfway along its length, it would form two Moebius strips!

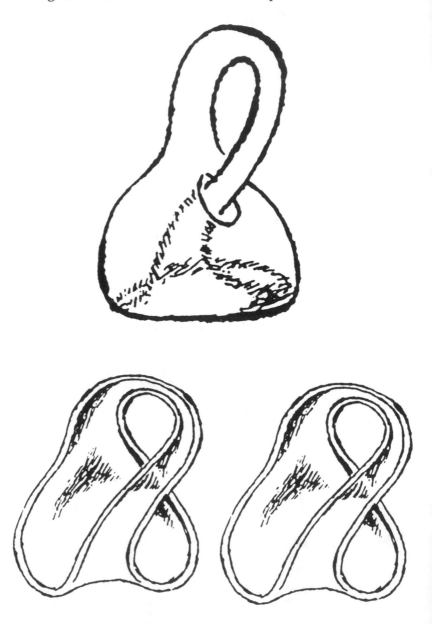

This puzzle was created by the famous puzzlist Sam Loyd. The object is to work one's way out of the diamond. Start at the

A Sam Loyd Puzzle

center, where the #3 appears. This number indicates you must move 3 squares or spaces right, left, up, down, or diagonally. Doing this, you land on another number which tells you how many more square spaces you can move in one of the eight possible directions.

Good luck!

See the appendix for the solution to *Sam Loyd's puzzle*.

Mathematics & paperfolding

Most of us at one time or another have folded a piece of paper and tucked it away, but fewer of us have intentionally folded paper in order to study the mathematical ideas it revealed. Paper folding can be both educational and recreational. Even Lewis Carroll was a paper folding enthusiast. Although paper folding transcends many cultures, it is the Japanese who are associated with developing and popularizing it into the art form called origami.

Some mathematical aspects of paperfolding

Many geometric concepts naturally appear when paper folding. Some of these are: *square, rectangle, right triangle, congruent, diagonal, midpoint, inscribe, area, trapezoid, perpendicular bisector, the Pythagorean Theorem, geometric & algebraic ideas*

Here are some examples of paperfolding that illustrate the use of these concepts.

i) from a **rectangular** shaped paper form a **square**

 cut-off this part

ii) with the square paper form 4 **congruent right triangles**

iii) find the **midpoint** of a side of a square

iv) **inscribe** a square in the paper square

 or

v) studying the paper's creases, notice that the inscribed square is 1/2 the **area** of the large square

vi) form two congruent **trapezoids** by taking a square sheet of paper and folding it along any edge so the crease passes through the center

vii) make the **perpendicular bisector** of a segment by folding the square in half–the crease will be the perpendicular bisector to the side of the square

viii) Demonstrate the **Pythagorean Theorem**
Fold the square paper as shown in the diagram.

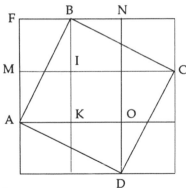

c^2 = area of square ABCD
a^2 = area of square FBIM
b^2= area of square AFNO

By matching up congruent shapes,
the area of square FBIM = area of $\triangle$ ABK
> *and*
the area of AFNO = the area of BCDAK
(the remaining area of square ABCD).

Thus, $a^2 + b^2 = c^2$

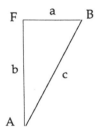

ix) Demonstrate the **theorem that the measure of the angles of a triangle total 180 degrees,** by taking any shaped triangle and folding it along the dotted lines, as illustrated.

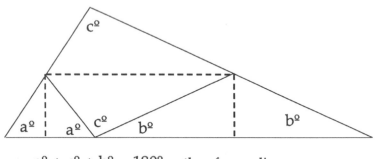

$$a^{\circ} + c^{\circ} + b^{\circ} = 180^{\circ}$$ —they form a line

x) Construct a **parabola** by folding **tangent lines.**

procedure: Locate the focus point of the parabola a few inches from the side of a sheet of paper. Crease the paper between 20 and 30 times as illustrated in the diagram. These creases are the tangents of the parabola and outline the curve.

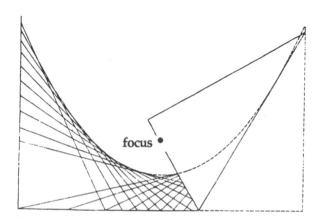

focus

E ach term in the Fibonacci sequence[1] is generated by adding the two previous terms. Any sequence generated in this manner is called a Fibonacci-like sequence.

The Fibonacci Trick

1, 1, 2, 3, 5, 8, 13, 21, 34, 55, 89, 144, 233, 377, 610, 987, 1597, 2584, 4181, 6765, 10946, 17711, 28657, 46368, 75025, 121393, 196418, 317811, 514229, ...

Select any two numbers and generate a Fibonacci-like sequence starting with the two numbers you selected. The sum of the first ten numbers in your sequence will automatically be eleven times the seventh term.

Can you prove this for any two starting numbers?

See appendix for proof of *Fibonacci trick*.

---

[1]See section on **The Fibonacci sequence** for more information.

The Evolution of Mathematical Symbols

From the early Babylonian scribe poking his cuneiform into a clay tablet to take the space allotted for zero, mathematicians have devised symbols for concepts and operations — for clarity and certainly to save time, effort and space.

During the fifteenth century, some of the first symbols used for *plus* and *minus* were *p* and *m*, respectively. German merchants, for example, were using + and - signs to denote defective weights. At this time + and - signs were adopted by mathematicians, and after 1481 they began to appear in manuscripts. The symbol for *times* , x, is attributed to William Oughtred (1574-1660), but it met opposition from some mathematicians because they contended it would be confused with the letter x.

There were often as many symbols for a concept as there were mathematicians. For example, in the sixteenth century Francois Vieta first used the word *aequalis* and later the symbol, ~, to denote equality. Descartes preferred ∝ for equality, but it was Robert Recorde's (1557) symbol, =, which ultimately was adopted. For him, parallel lines were objects that were most alike and denoted equality.

Although letters had been used for unknowns by the ancient Greek mathematicians Euclid and Aristotle, this was not a common practice. In the 16th century, words such as *radix* (Latin for root), *res* (Latin for thing), *cosa* (Italian for thing), *coss* (German for thing) were used to refer to unknowns. Between the years 1584-1589 when lawyer Francois Vieta was between terms in the Brittany Parliament, he made an extensive study of the works of many mathematicians. He developed the notion of using letters for positive known and unknown quantities. Descartes modified his idea and introduced the notion of using the first letters of the alphabet for known quantities and the last

This symbol was first used by Italian mathematician Fibonacci in 1220. It denotes √ and probably was derived from the Latin word radix meaning root. The symbol we use today, √ , is from 16th century Germany.

The German mathematician, Christoff Rudolff used this symbol for cube root in 1525. ∛ is from France in the 17th century.

17th century German mathematician, Leibniz, chose this symbol for multiplication.

This reversed D was used to denote division by Frenchman J.E.Gallimard in the 1700's.

In 1859 Benjamin Peirce, a Harvard professor, used this symbol for pi. π originated in England in the 18th century.

This symbol was used for addition by Tartaglia, Renaissance mathematician. It is derived from the Italian word piu (more).

The ancient Greek mathematician, Diophantus used this symbol for subtraction.

letters for unknown quantities. And finally, in 1657, letters were used for both positive and negative numbers by John Hudde.

∞ had been used by the Romans to denote 1000, and later any very large number. In 1655, John Wallis, an Oxford professor, used the symbol, ∞, for infinity for the first time. But the symbol was not generally employed until 1713 when it was used by Bernoulli.

Other symbols that evolved were the use of parentheses in 1544, square root brackets and braces in 1593 and the square root radical, which was devised by Descartes (he used $\sqrt{c}$ for cube root).

It is hard to imagine working mathematical problems without a + sign or symbol for 0, or any of the other mathematical notations we take for granted. It is also difficult to realize it has taken centuries for them to evolve and be universally accepted.

The chart compares mathematical symbols and phrases used in the past and present.		
	PAST	*PRESENT*
	℞	$\sqrt{\ }$
	p	+
	m	-
	v	*used under the radical*
Cardan(1501-1576)	℞.v.7.p:℞.14	$\sqrt{7}+\sqrt{14}$
Chuquet 1484	$12^3 + 12^0 + 7^{1m}$	$12x^3 + 12 + 7x^{-1}$
Bombelli	③	x^3
Stevin 1585	1⓪+3①+ 6②+ ③	$1+3x+6x^2+x^3$
	①/2	$\sqrt{\ }$
	①/3	$\sqrt[3]{\ }$
Descartes	$1+3x+6xx+x^3$	$1+3x+6x^2+x^3$

This sketch by Leonardo da Vinci shows his use of regular polygons in the design of a church. Leonardo's interest and study in geometric structures

Some geometric designs of Leonardo da Vinci

and his knowledge of symmetry were his tools for creating architectural plans for adding chapels to churches without upsetting the design and symmetry of the main building.

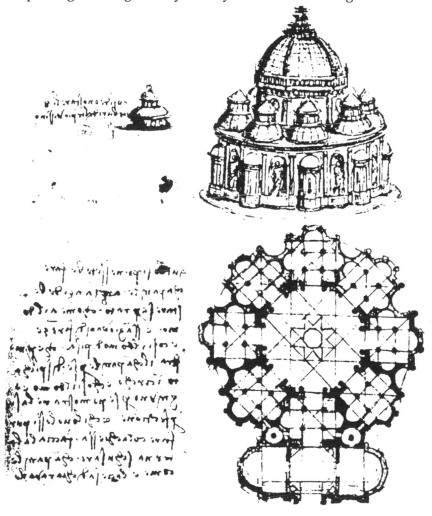

Ten Historical Dates

Written below in different number systems are ten historical dates. Write each in base ten, and identify what historical event took place in that year.

This chart of various number system may help you in deciphering the numbers.

Today	1	2	3	4	5	6	7	8	9	10	11	12	13	14	15	16
Babylonian (1500 B.C.)	Y	YY	YYY	YYYY	YYY YY	YYY YYY	YYY YYY Y	YYY YYY YY	YYY YYY YYY	⟨	⟨Y	⟨YY	⟨YYY	?	?	?
Chinese (500 B.C.)	一	二	三	四	五	六	七	八	九	十	土	土	圭	?	?	?
Greek (400 B.C.)	A	B	Γ	Δ	E	F	Z	H	Θ	I	IA	IB	IΓ	?	?	?
Egyptian (300 B.C.)	I	II	III	IIII	III II	III III	IIII III	IIII IIII	IIII IIII I	∩	∩I	∩II	∩III	?	?	?
Roman (200 B.C.)	I	II	III	IV	V	VI	VII	VIII	IX	X	XI	XII	XIII	?	?	?
Maya (300 A.D.)	•	••	•••	••••	—	•/—	••/—	•••/—	••••/—	=	•/=	••/=	•••/=	?	?	?
Hindu (11th century)	?	?	?	?	?	?	?	?	?	?0	??	??	??	?	?	?
Base Two (computers)	1	10	11	100	101	110	111	1000	1001	1010	1011	1100	1101	?	?	?

See appendix for solution.

Napoleon's Theorem

Napoleon Bonaparte (1769-1821) had a special respect for mathematics and mathematicians, and enjoyed the subject himself. In fact, he is attributed with the following theorem.

If three equilateral triangles are constructed off the sides of any triangle, then the centers of the circles which circumscribe each equilateral triangle are vertices of another equilateral triangle.

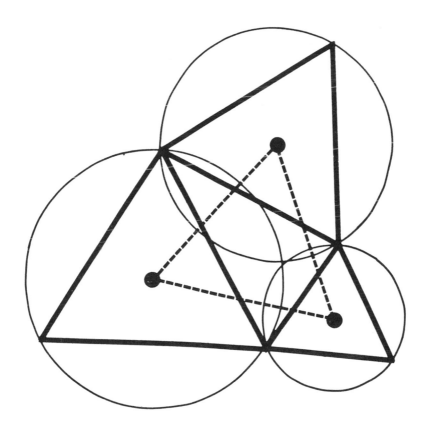

Lewis Carroll – The Mathematician

Charles Lutwidge Dodgson(1832-1898), an English mathematician and logician, is better known by his pseudonym Lewis Carroll and as the author of *Alice in Wonderland* and *Alice through the Looking Glass*. In addition, he published many texts dealing with various fields of mathematics. His book *Pillow Problems* – 72 problems which were nearly all written and solved while he lay awake at night – deals with arithmetic, algebra, geometry, trigonometry, analytical geometry, calculus and transcendental probability.

"Contrariwise," continued Tweedledee "if it was so, it might be; and if it were so, it would be: but as it isn't, it ain't. That's **logic**." —Lewis Carroll

A *Tangled Tale* was originally printed as a monthly magazine article, and later compiled into a delightful story containing mathematical puzzles in its ten chapters. It is said that Queen Victoria was so taken by Carroll's Alice books that she sent for every book he had written. Imagine her surprise at the stack of math books she received.

problem # 8 from *Pillow Problems*

"Some men sat in a circle, so that each had 2 neighbours; and each had a certain number of shillings. The first had one shilling more than the second, who had one shilling more than the third, and so on. The first gave one shilling to the second, who gave two shillings to the third, and so on, each giving one shilling more than he received, as long as possible. There were then 2 neighbours, one of whom had 4 times as much as the other. How many men were there? And how much had the poorest man at first?"

See the appendix for the solution.

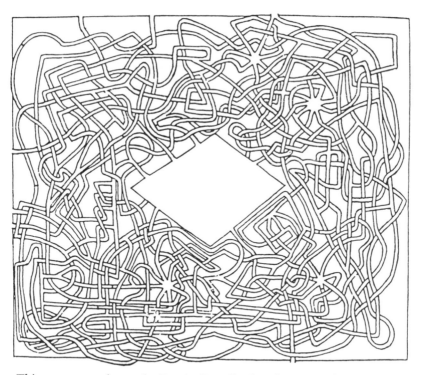

This maze was drawn by Lewis Carroll when he was in his twenties. He made the maze with paths that cross over and under each other. The object is to find a path out of the center.

Counting on Fingers

Because paper writing materials were at a premium during the Middle Ages, counting and communicating results by finger signs were often used. As the diagram illustrates the finger system was able to represent both small and large numbers.

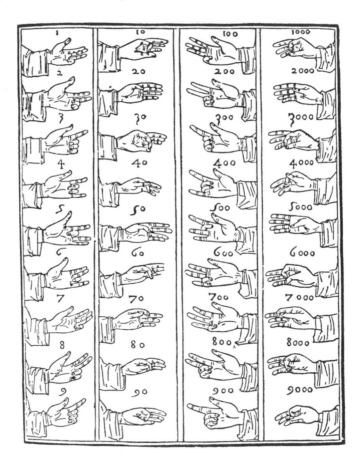

A Twist to the Moebius Strip

The figures below involve the use of the Moebius strip. If you make paper models of these topological models, and cut them along the dotted lines, one will become a square and the other will come apart in some way.

Heron's Theorem

Many have learned in geometry how to compute the area of a triangle using the length of its altitude and of its related base. But without Heron's Theorem, calculating its area knowing only the lengths of its three sides would require knowledge of trigonometry.

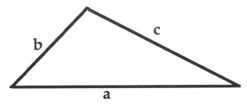

Heron is best known in the history of mathematics for the formula,

$$\text{area of triangle} = \sqrt{s(s-a)(s-b)(s-c)}$$

where a, b, and c are the lengths of the sides of the triangle and s is half the sum of the lengths of the sides of the triangle.

Heron's formula was known earlier to Archimedes, who probably had proven it, but the earliest written record we have is from Heron's writing, *Metrica*. Heron is best described as a non-classical type of mathematician. He was more concerned about the practicality of mathematics than about theory and the treatment of mathematics as a science and an art. As a result, he is also remembered as the inventor of a primitive type of steam engine, various toys, a fire engine that pumped water, an altar fire that lit as soon as the temple doors opened, a wind organ, and various other mechanical devices based on the properties of fluids and the laws of simple mechanics.

These rare Gothic plans show the use of geometry and symmetry in the architecture of the *Dome of Milan*. These plans were published in 1521 by Caesar Caesariano, master architect of the *Dome of Milan*.

A look at Gothic architecture & geometry

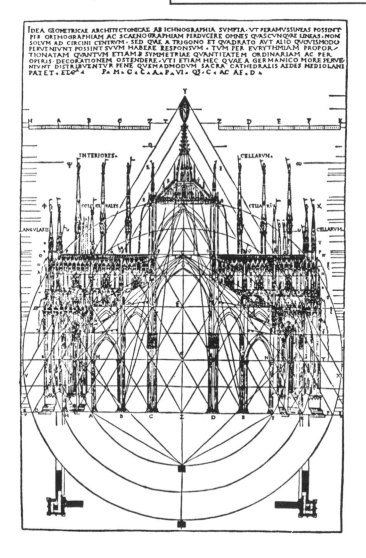

IDEA GEOMETRICAE ARCHITECTONICAE AB ICHNOGRAPHIA SVMPTA. VT PERAMVSSIM EAS POSSINT PER ORTHOGRAPHIAM AC SCAENOGRAPHIAM PERDVCERE OMNES QVASCVNQVAE LINEAS. NON SOLVM AD CIRCINI CENTRVM. SED QVAE A TRIGONO ET QVADRATO AVT ALIO QVOVISMODO PERVENIVNT POSSINT SVVM HABERE RESPONSVM. TVM PER EVRYTHMIAM PROPOR- TIONATAM QVANTVM ETIAM Đ SYMMETRIAE QVANTITATEM ORDINARIAM AC PER OPERIS DECORATIONEM OSTENDERE. VTI ETIAM HEC QVAE A GERMANICO MORE PERVE- NIVNT DISTRIBVENTVR PENE QVEMADMODVM SACRA CATHEDRALIS AEDES MEDIOLANI PATET. ETC▵ P▪M▪C▪C▪A▪P▪VI▪Q§▪C◂AC AF▪D ▵

Napier's Bones

Working with complicated figures had become more and more tedious, especially for scientists doing astronomical calculations, seamen needing to perform practical navigation problems and merchants rendering their accounts. Then, in the 17th century, John Napier (1550–1617), a famous Scottish mathematician, revolutionalized computation with his discovery of logarithms *(a method of using exponents to perform complicated multiplication and division by converting it to addition and subtraction).* [1] Napier's method of computation with logarithms and the tables he developed simplified difficult calculations involving multiplication, division, exponents, and finding roots. Although the theory of logarithmic and exponential functions is an essential part of mathematics, with the introduction of modern electronic calculators and computers, the logarithmic tables and their uses have become as obsolete as the slide rule. But the development of these tables and shortcut methods of computation were a marvelous computing method widely used for centuries by mathematicians, accountants, navigators, astronomers and scientists.

Using logarithms, Napier also invented numbered rods, called **Napier's bones,** to help merchants' accounting. Merchants would carry a set of rods, made of ivory or wood, with them to perform multiplication, division, square roots and cube roots. Each rod was a multiplication table for the digit at its top. To multiply 298 by 7 one would line up the 2, 9, and 8 rods and then count down to the seventh row, and sum the two numbers as illustrated.

---

[1] For example, to perform the operation—3600/.072, one would convert these numbers to their exponent (i.e. logarithmic) form, using log tables. To perform division of numbers written in exponent form of the same base, is a simple matter of subtracting their exponents. So the two logarithmic numbers for 3600 and .072 are subtracted, and the result is converted back by using log tables to a number in base ten.

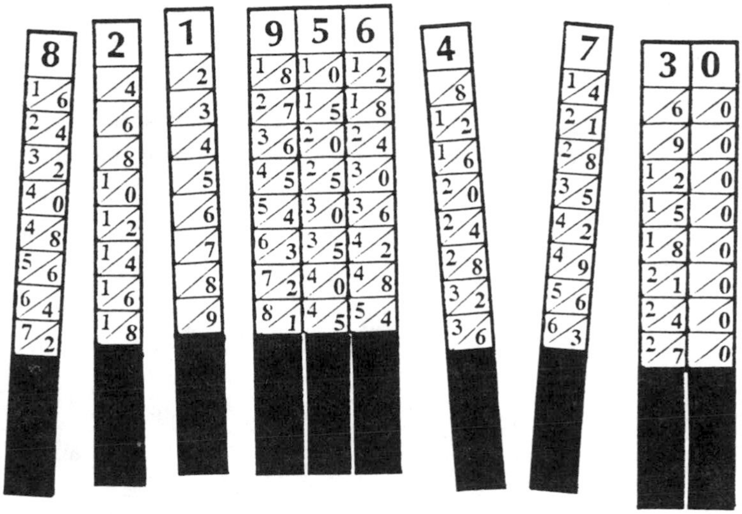

$$298$$
$$\times 7$$
$$\overline{2086}$$

$$165$$
$$+\ 436$$
$$\overline{2086}$$

Art & Projective Geometry

Consciously or unconsciously mathematics has influenced art and artists over the centuries. Projective geometry; the golden mean; proportion; ratios; optical illusions; symmetries; geometric shapes; designs and patterns; limits and infinity; and computer science are some of the areas of mathematics which have affected various facets and periods of art — be it primitive, classical, Renaissance, modern, pop or art deco.

The superimposed lines illustrate the use of projective geometry by Leonardo da Vinci in his masterpiece, *The Last Supper*.

An artist painting a three dimensional scene on a two dimensional canvas must decide how things change when viewed from different distances and positions. It is in this area that projective geometry developed and played a major part in Renaissance art. *Projective geometry is a field of mathematics that deals with properties and spatial relations of the figures as they are projected — and therefore with problems of perspective.* To create their realistic three dimensional paintings, Renaissance artists used the now established concepts of projective geometry — point of projection, parallel converging lines, vanishing point. Projective geometry was one of the first non-Euclidean geometries to surface. These artists wanted to depict realism. They reasoned that if they perceived a scene outside through a window, it would be possible to project what they saw onto the window as a collection of points of vision if they kept their eye at a single focal point. Thus, the window could act as their canvas. Various devices were created to enable the artist to actually transform the window into a canvas. These wood cuts by Albrecht Dürer illustrate two such devices. Note that the artist's eye is at a fixed point.

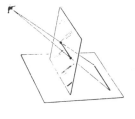

Infinity & the Circle

Every circle has a fixed circumference — a length of a finite amount. One method for deriving the formula for the circumference of a circle is to use the concept of infinity. Study the sequence of perimeters of inscribed regular polygons. (A regular polygon has all sides the same size and all angles the same measure.) By calculating and studying the perimeter of each inscribed regular polygon, one notices as the sides of the polygon increase its perimeter comes closer and closer to the circumference of the circle. In fact, the limit of the perimeters as the number of sides approach infinity is the circumference of the circle. The diagram illustrates that the more sides a polygon has the closer its sides are to the circle, and the more it resembles a circle.

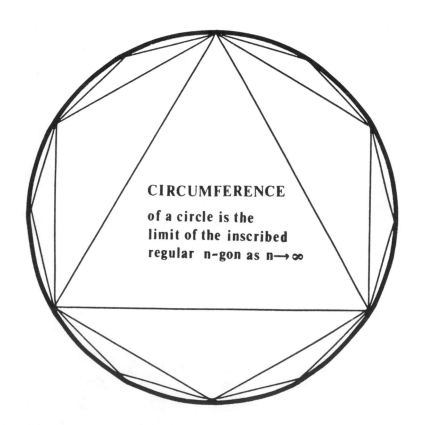

CIRCUMFERENCE

of a circle is the
limit of the inscribed
regular n-gon as n→∞

Take any size circular track made up of any two concentric circles, as illustrated below. Can you show that the area of the

The Amazing Track

track is equal to the area of a circle whose diameter is a chord of the larger circle but tangent to the smaller circle?

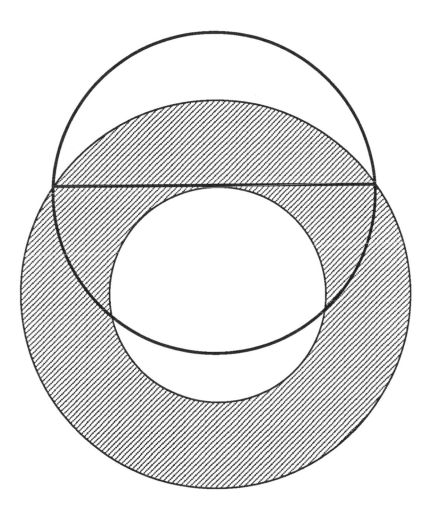

See appendix for proof.

The Persian Horses & Sam Loyd's Puzzle

This 17th century Persian drawing ingeniously pictures four horses. *Can you locate the four horses?*

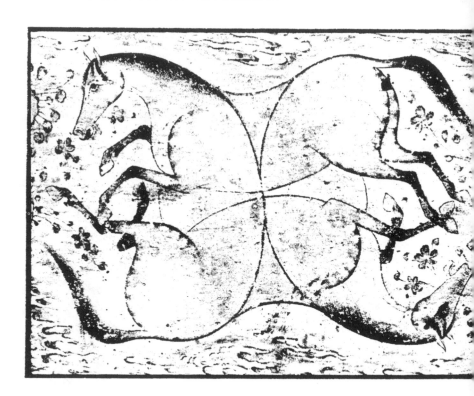

See appendix for solution.

This drawing may have been the inspiration for **the jockey and donkey puzzle** by master puzzlist Sam Loyd (1841-1911).

Loyd's original version was created around 1858, when he was a teenager.

The problem is to cut the picture into three rectangles along the dotted lines, and rearrange the rectangles without folding them, to show two jockeys riding two galloping donkeys.

This puzzle was an instant success. In fact, it was so popular that Sam Loyd reportedly earned $10,000 in a matter of weeks.

See appendix for solution to *Sam Loyd's puzzle.*

Lunes

The word **lune** has its origin from the Latin word *lunar* — moon shaped. Lunes are plane regions bounded by arcs of different circles (see shaped crescents in the diagram). Hippocrates of Chios (460-380 B.C.), not to be confused with the physician of Cos (author of the Hippocratic oath), did extensive study with lunes. He probably believed that they could somehow be used to solve the problem of squaring a circle.[1]

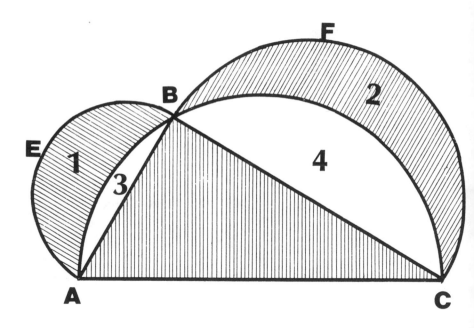

He found and proved that:

> *The sum of the areas of two lunes constructed*
> *on two sides of a triangle which is inscribed*
> *in a semicircle, equals the area of that triangle.*

---

[1] See section on the **Impossible Trio**.

If $\overset{\frown}{ABC}$, $\overset{\frown}{AEB}$, $\overset{\frown}{BFC}$ are semicircles, then
the area of lune(1) +area of lune(2)=area of triangle ABC.

proof

area of semi⊙$\overset{\frown}{AEB}$/areaofsemi⊙$\overset{\frown}{ABC}$=$(\pi|AB|^2/8)/(\pi|AC|^2)/8$

$$=|AB|^2/|AC|^2 —$$

area of semi⊙$\overset{\frown}{AEB}$ = area of semi⊙ $\overset{\frown}{ABC}(|AB|^2/|AC|^2)$

—*line (a)*

similarly

area of semi⊙$\overset{\frown}{BFC}$ = areaof semi⊙$\overset{\frown}{ABC}$ $(|BC|^2/|AC|^2)$ —*line(b)*

Now adding like terms and factoring lines (a) and (b) —
area of semi⊙$\overset{\frown}{AEB}$ + area of semi⊙$\overset{\frown}{BFC}$ =

areaof semi⊙$\overset{\frown}{ABC}$ $(|AB|^2 + |BC|^2)/|AC|^2$ –*line (c)*

ΔABC is a right triangle because it is inscribed on a semi⊙.
Thus,

$$|AB|^2 + |BC|^2 = |AC|^2$$ —*by the Pythagorean theorem*

Substituting this in line (c) and simplying we get —
area of semi⊙$\overset{\frown}{AEB}$ + area of semi⊙$\overset{\frown}{BFC}$ = area of semi⊙$\overset{\frown}{ABC}$
subtracting   area (3) +   area (4)   =area (3) + area (4)

QED   area of lune(1) + area of lune (2) = area Δ ABC

Although Hippocrates was not successful in his effort to square a circle, his quest helped him discover many new important mathematical ideas.

Hexagons in Nature

Many of nature's creations are beautiful models for such mathematical objects as squares and circles. The regular hexagon is one of the geometric figures found in nature. A hexagon is a 6-sided figure. It is a regular hexagon if all its sides are the same length and its angles are the same size.

Mathematicians have shown that only regular hexagons, squares and equilateral triangles can be fitted together (tessellated) so there is no wasted space.

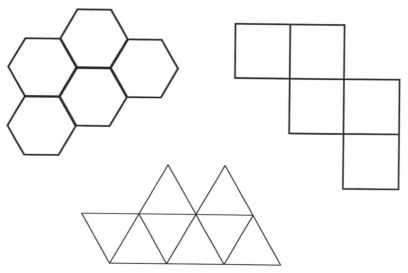

Of the three, the hexagon has the smallest perimeter when their areas are equal. This means, that in forming a hexagonal cell to build the honeycomb, the bee uses less wax and does less work for the same space. The hexagon shape is found in honeycombs, snowflakes, molecules, crystals, sealife and other forms.

Walking in a snow flurry you are in the midst of some wondrous geometric shapes. The snowflake is one of the most exciting examples of hexagonal symmetry in nature. One can spot the hexagons in the formation of each snowflake. The infinite number of combinations of hexagonal patterns, explains the common belief held that no two snowflakes are alike.[1]

[1]Nancy C. Knight of the National Center for Atmospheric Research in Boulder Colorado, may have discovered the first set of identical snowflakes. They were collected on November 1, 1986.

Googol & Googolplex

A googol is 1 followed by a hundred zeros, 10^{100}. The name *googol* was coined by mathematics author Dr. Edward Kasner's nine year old nephew. The nephew also suggested another number much larger than a googol called a *googolplex,* which he described as 1 followed by as many zeros as you could write before your hand got tired. The mathematical definition for a googolplex is 1 followed by a googol of zeros, 10^{googol}.

10 000 000 000 000 000 000 000 000 000 000 000 000 000 000 000 000 000 000 000 000 000 000 000 000 000 000 000 000 000 000 000 000 000 000 000 000 000 000 000 000 000 000 000

Using large numbers:

1) If the entire universe were filled with protons and electrons so that no vacant space remained, the total number would be 10^{110}. This number is larger than a googol but much less than a googolplex.

2) The number of grains of sand on Coney Island is about 10^{20}.

3) The number of words printed since the Guttenberg Bible (1456) until the 1940's is about 10^{16}.

This 3 by 3 magic cube is an arrangement of the first 27 natural numbers so that the sum of each row or column of three numbers is 42.

A Magic Cube

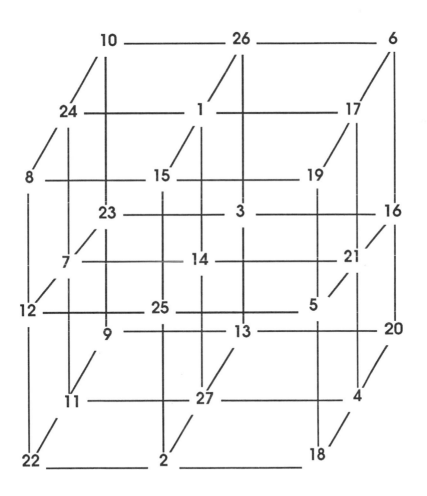

Fractals— real or imaginary?

For many centuries the objects and concepts of Euclidean geometry (such as point, line, plane, space, squares, circles, ...) were considered to be the things that describe the world in which we live. The discovery of non-Euclidean geometries has introduced new objects that depict the phenomena of the universe. Fractals are such objects. These shapes are now considered to portray objects and phenomena of nature. The idea of fractals originated in the work of mathematicians from 1875 to 1925. These objects were labeled *monsters,* and believed to have little scientific value. Today they are known as *fractals* — coined by Benoit Mandelbrot in 1975, who made extensive discoveries in this area. Technically, a *fractal is an object whose detail is not lost as it is magnified* — in fact the structure looks the same as the original.

The snowflake curve[1] is an example of a fractal generated by equilateral triangles being added to sides of existing equilateral triangles.

In contrast, a circle appears to become straight as a portion of it is magnified. However, there are two categories of fractals — the *geometric fractals* which continuously repeat an identical pattern and the *random fractals.* Computers and computer graphics are responsible for bringing these "monsters" back to life by almost instantaneously generating the fractals on the computer screen and thereby displaying their bizarre shapes, artistic designs or detailed landscapes and scenes.

It had been felt that the orderly shapes of Euclidean geometry were the only ones applicable to science, but with these new

---

[1]See section on snowflake curve for additional information.

forms nature can be viewed from a different perspective. Fractals form a new field of mathematics — sometimes referred to as the geometry of nature because these strange and chaotic shapes describe natural phenomena such as earthquakes, trees, bark, ginger root, coastlines, and have applications in astronomy, economics, meteorology and cinematography.

The Peano curve is an example of a fractal and also a space-filling curve. In a space filling curve, every point within a given area is traced and gradually blackens the space. This example is partially traced.

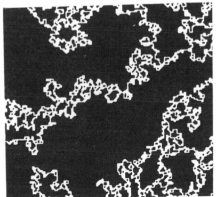

a random fractal

Cesaro curve – a fractal

Nanoseconds – measuring time on computers

It takes an electric impulse one-billionth of a second to travel 8 inches. One-billionth of a second has come to be called a **nanosecond**. Light travels one foot in one nanosecond. Computers today are built to perform millions of operations per second. To get a feeling for how fast a large computer can work, consider half of a second. Within a half a second the computer would have performed the following tasks:

1) *debited 200 checks to 300 different bank accounts;*
2) *examined the electrocardiograms of 100 patients;*
3) *scored 150,000 answers on 3000 exams, and evaluated the validity of each question;*
4) *figured the payroll for a company with one-thousand employees;*
5) *plus time left over for some other tasks.*

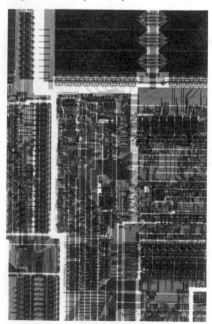

It's mind boggling to try to imagine how fast a computer would be if it could be powered by light rather than electricity. What kind of number system would be needed to harness the use of light? Would it be based on the number of colors in the spectrum of light? Or perhaps some other property of light?

Leonardo da Vinci was intrigued by many fields of study and their interrelation. Mathematics was one of them, and he used many concepts of

Geodesic Dome of Leonardo da Vinci

mathematics in his art, his architectural designs and his inventions. Here is a rendition of a geodesic dome he sketched.

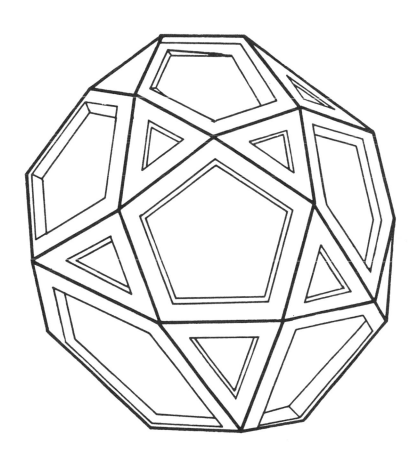

Magic Squares

Magic squares have intrigued people for centuries. From ancient times they were connected with the supernatural and the magical world. Archeological excavations have turned them up in ancient Asian cities. In fact, the earliest record of the appearance of a magic square is about 2200 B.C. in China. It was called *lo-shu*. Legend has it that this magic square was first seen by Emperor Yu on the back of a divine tortoise on the bank of the Yellow River.

Lo-Shu

The black knots represent even numbers and the white knots are odd numbers. In this magic square, its magic number (the total of any row, column or diagonal) is fifteen.

In the Western world magic squares were first mentioned in 130 A.D. in the work of Theon of Smyrna. In the 9th century, magic squares crept into the world of astrology with Arab astrologers using them in horoscope calculations. Finally, with the works of the Greek mathematician Moschopoulos in 1300 A.D., magic squares and their properties spread to the western hemisphere (especially during the Renaissance period).

SOME PROPERTIES OF MAGIC SQUARES:

The **order** of a magic square is defined by the number of rows or columns. For example, this magic square has order 3 because it has 3 rows.

16	2	12
6	10	14
8	18	4

The **magic** of a magic square arises from all the fascinating properties it possesses. Some of the properties are:

1) Each row, column, and diagonal total the same number. This magic constant can be obtained in one of the following ways:

    a) Take the magic square's order, n, and find the value of $1/2(n(n^2+1))$ where the magic square is composed of the natural numbers $1,2,3,\ldots,n^2$.

8	1	6
3	5	7
4	9	2

order 3.
magic number
$= 1/2(3(3^2+1))=15$

15

1	2	3
4	5	6
7	8	9

15

    b) Take any size magic square and, starting from the left hand corner, place the numbers sequentially along each row. The sum of the numbers in either diagonal will be the magic constant.

2) Any two numbers (in a row, column, or diagonal) that are equidistant from the center are complements. Numbers of a magic square are complements if their sum is the same as the sum of the smallest and largest numbers of that magic square.

8	1	6
3	5	7
4	9	2

This magic square has complements—
8 & 2   6 & 4   3 & 7   1 & 9

WAYS TO TRANSFORM AN EXISTING MAGIC SQUARE TO ANOTHER MAGIC SQUARE:

3) Any number may be either added or multiplied to every number of a magic square and it will remain a magic square.

4) If two rows and two columns, equidistant from the center, are interchanged the resulting square is a magic square.

5) a) Interchanging quadrants in an even order magic square results in a magic square.

a quadrant is one of the four sections of a magic square

b) Interchanging partial quadrants and rows in an odd order magic square results in a magic square.

More has been written on magic squares than any other mathematical recreation. Benjamin Franklin spent considerable time devising methods of forming magic squares. It is quite challenging to take the first 25 counting numbers and arrange them in a 5x5 square so that each row, column, and

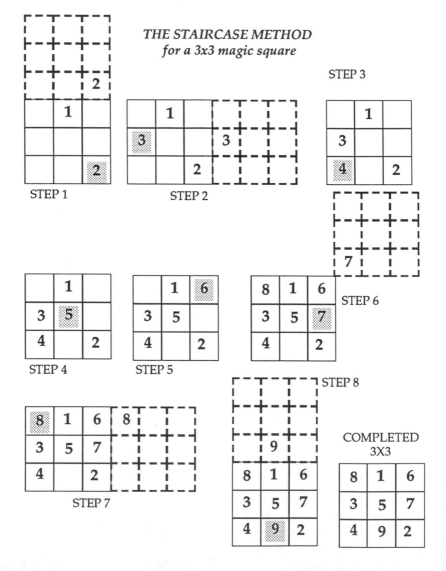

THE STAIRCASE METHOD
for a 3x3 magic square

diagonal total the same amount. This would be referred to as a magic square of order 5. Any magic square with an odd number of rows or columns is an odd order magic square. If there are an even number, it is an even order magic square. A general method for making any size even order magic square is still being explored. On the other hand, for the odd order ones there are a number of general methods that can be used to make any size odd order magic square. The staircase method, invented by La Loubere is well known among magic square enthusiasts. The diagrams illustrate a 3x3 magic square being made using this method.

THE STAIRCASE METHOD

1) Start with the number 1 in the middle box of the top row.

2) The next number is placed diagonally upward in the next box, unless it is occupied. If it lands in a box in an imaginary square outside your magic square, find its location in your magic square by matching the location its holds in the imaginary square to your magic square.

3) If in your magic square the diagonally upward box is occupied, then place the number in the box immediately below the original number, e.g. for numbers 4 and 7.

4) Continue following steps (2) and (3) to obtain the locations of the remaining numbers for the magic square.

Now try making the 5x5 magic square using the first 25 counting numbers and the staircase method. Test some of the methods for altering magic squares on it to see how they work.

Using any of the magic squares you have constructed, multiply each of its numbers by a constant you choose. Is the resulting square a magic square?

For the even order magic squares various methods have been devised for constructing a particular even order magic square.

example: The **diagonal method** applies only to a 4x4 magic square.

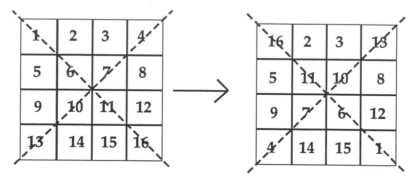

procedure:
Begin by sequentially placing the numbers in the rows of the magic square. If a number lands on one of the diagonals, its location must be switched with its complement.

With a 4x4 magic square, either rows or columns may be interchanged, and it will remain a magic square. Also if quadrants are interchanged, it results in a magic square.

See if you can devise your own methods for making magic squares of other even orders, or discover a general method for all even order magic squares.[1] You may also want to research other methods for making any odd order magic squares that have been devised.

---

[1] Many have devoted much time and effort to discovering a general method of even order magic squares. Hyman Sirchuck of Howell, New Jersey claims to have devised a method for making even order magic squares.

The Fibonacci sequence is 1,1,2,3,5,8,13,... in which each term is the sum of the two terms preceding it. When the Fibonacci numbers 3, 5,

The "Special" Magic Square

8, 13, 21, 34, 55, 89, 144 are matched up with counting numbers 1, 2, 3, 4, 5, 6, 7, 8, 9, a new square is formed. It does not have the traditional properties of a magic square, but the sum of the products of the 3 rows (9078+9240+9360 =27,678) equals the sum of the products of the 3 columns (9256+9072+9350=27,678).

8	1	6
3	5	7
4	9	2

89	3	34
8	21	55
13	144	5

Chinese Triangle

Mathematics is universal. History shows its uses and discoveries are not unique to one region as illustrated by this Chinese version of the Pascal triangle. Although Pascal made some significant discoveries in his triangle of numbers, it appeared in a book printed in around 1303, three-hundred and two years before Pascal was born![1]

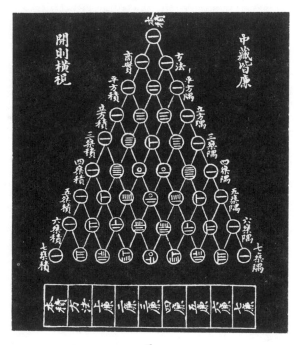

```
              1
           1     1
        1     2     1
     1     3     3     1
  1     4     6     4     1
  .   .   .   .   .   .   .
```

---

[1]See section on **Pascal's triangle.**

Archimedes of Syracuse (287 B.C. to 212 B.C.) was a leading mathematician of the Hellenistic Age.

The Death of Archimedes

During the Second Punic War, Syracuse was besieged by the Romans from 214 B.C. to 212 B.C. At this time Archimedes invented ingenious defense weapons — catapults, pulleys and hooks to raise and smash Roman ships, parabolic mirrors to set fire to ships — which were responsible for holding the Romans back for nearly three years. Syracuse eventually fell to the Romans, however. Marcus Claudius Marcellus, commander, gave express orders that Archimedes was not to be harmed. A Roman soldier entered the home of Archimedes and found him working on a math problem, totally oblivious to his presence. The soldier ordered him to stop, but Archimedes did not pay any attention. Angered, the soldier killed Archimedes with his sword.

A Non-Euclidean World

The 19th century was a period of revolutionary ideas in politics, art, and science, as well as in mathematics with the development of its non-Euclidean geometries. Their discoveries marked the beginning of modern mathematics in the same way impressionistic painting marked the beginning of modern art.

During this period, hyperbolic geometry (one of the non-Euclidean geometries) was discovered independently by Russian mathematician Nicolai Lobachevsky (1793-1856) and Hungarian mathematician Johann Bolyai (1802-1860).

An abstract design of Poincaré hyperbolic geometric model.

We find that hyperbolic geometry, like other non-Euclidean geometries, describes properties that are alien, since we are conditioned to think of geometry in Euclidean terms. For example, in hyperbolic geometry, *line* does not imply straight and *parallel lines* do not remain equidistant though they do not intersect because they are asymptotic. When non-Euclidean geometries are studied in detail one finds that they may indeed give a more accurate description of phenomena in our universe. As a result, *different worlds* have been described in which these geometries could exist.

One such world is the model created by the French mathematician Henri Poincaré (1854 – 1912). His imaginary universe is bounded by a circle (visualize a sphere for a 3-dimensional model) whose temperature at the center is absolute 0. As one travels from the center the temperature around rises. Assume the objects and the inhabitants in this universe are unaware of changes in temperature, but the size of everything changes as it moves about. In fact, every object and living thing enlarges as it approaches the center and shrinks proportionally as it approaches the boundary. Since *everything* changes size, one would be unaware and unable to detect size changes. This means one's steps would become smaller as one moves toward the boundary, and indeed, it would appear that one is no nearer the boundary than before approaching it. This phenomenon makes this world appear infi-

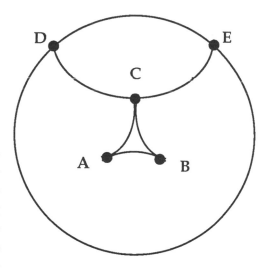

nite, and here the shortest distance between two points is a curved line because to get from A to B it would take less steps (since they are larger) if we move toward the center in an arc shape. Here is a world in which a triangle's sides would be made up of arcs, as triangle ABC in the diagram. And even parallel lines have a new look. Line DCE is parallel to line AB since they would never intersect.

Actually Poincaré's universe may describe the world in which we live. If we look at our position in the universe and if we were able to travel distances measured in light years, then perhaps we would discover changes in our physical size. In fact in Einstein's theory of relativity the length of a ruler shortens as it approaches the speed of light!

Henri Poincaré

Poincaré was an *original thinker.* The diversity of subjects he lectured on while a professor at the Sorbonne in Paris (from 1881–1912) illustrate this point. His work and his ideas cover such subjects as electricity, potential theory, hydrodynamics, thermodynamics, probabilty, celestial mechanics, divergent series, asymptotic expansions, integral invariants, stability of orbits, shapes of celestial bodies, and the list goes on. His work can definitely be said to have stimulated mathematical thought of the twentieth century.

S_{quare} numbers, pyramidal numbers and their summations can be used in determining the number of cannon balls in a square based pyramid.

Cannon Balls & Pyramids

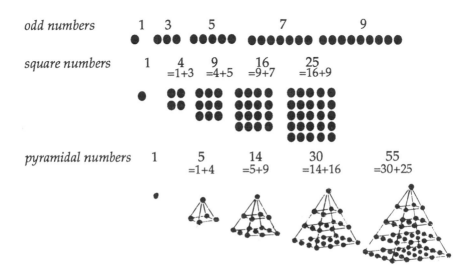

odd numbers   1   3   5   7   9

square numbers   1   4 =1+3   9 =4+5   16 =9+7   25 =16+9

pyramidal numbers   1   5 =1+4   14 =5+9   30 =14+16   55 =30+25

Study the patterns of these numbers.

How many cannon balls are there in the diagram below?

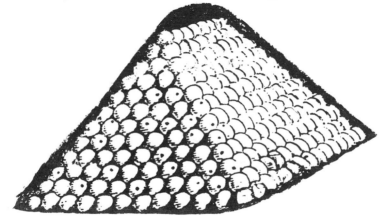

Conchoid of Nicomedes

Very often, searching for solutions to certain mathematical problems gives rise to new concepts and discoveries. The three famous construction problems of antiquity — *trisecting an angle* (dividing an angle into three congruent angles), *duplicating a cube* (constructing a cube with twice the volume of a given cube), and *squaring a circle* (constructing a square with the same area as a given circle) — stimulated mathematical thought, and as a result many ideas were discovered in an attempt to solve these. Although it has been shown that these three ancient problems are impossible to construct using only a *straightedge and compass*, they were solved using other means — one of which was the *conchoid.*

The *conchoid* is one of the curves of antiquity which was invented by *Nicomedes* (circa 200 B.C.) and used to duplicate a cube and trisect an angle.

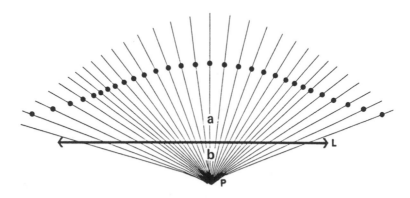

To form a conchoid begin with a line L and point P. Then draw rays through P which intersect L. Mark off a fixed distance, a, on each of these rays. The locus of these points form the conchoid. The curvature of the conchoid depends on the relationship between a and b, i.e. a=b, a<b, or a>b. The polar equation of the conchoid is r=a+bsecΘ .

To trisect <P, make <P one of the acute angles of right △QPR. Draw a conchoid with pole P and with $\overleftrightarrow{QR}$ as its fixed line, L. For the fixed distance from $\overline{QR}$ use 2h, twice the hypotenuse |PR|. At R construct $\overline{RS} \perp \overline{QR}$ intersecting the conchoid at S. Now <QPT can be shown to be one-third <QPR.

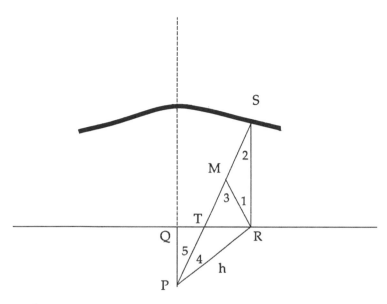

proof

Locate M, the midpoint of $\overline{TS}$, then |RM|=h, since △SRT is a right triangle and the midpoint of the hypotenuse is equidistant from its vertices.

Now m<1=m<2=k° because |MS|=|MR|=h. And m<3=2k°, since <3 is an exterior angle for △SMR. Also m<3=m<4=2k° because |MR|=|PR|=h.

m<5=m<2=k°, since $\overline{PQ} \parallel \overline{RS}$, because segments $\overline{PQ}$ and $\overline{SR}$ are coplanar and perpendicular to line $\overleftrightarrow{QR}$.

Thus, m<QPR=3k° and 1/3(m<QPR)=k°=m<5.

Therefore, <QPR is trisected.

The Trefoil Knot

Tieing knots has been a routine process for most of us from the time we mastered tieing our shoe-laces. Of course tieing knots can be an art, especially when one sees a sailor rigging a boat. But the subject of knots is also a mathematical idea in the field of topology. Knots form a relatively new area. The most important idea that has been proven about them thus far is that *a knot cannot exist in more than three dimensions.*

Making a trefoil knot

To form the *trefoil knot,* illustrated below, take a long strip of paper and give it 3 half twists. Now join its ends together with tape. Using a pair of scissors cut midway between its edges along the entire strip. You will end up with one band with a trefoil knot in it.

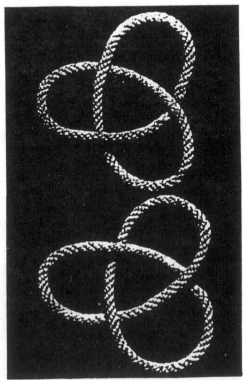

Benjamin Franklin's magic square features a variety of numerical oddities in addition to the regular properties of a magic square.[1] Each row totals 260. Halfway totals 130. A shaded diagonal up four numbers and down four totals 260.

The Magic Square of Benjamin Franklin

The sum of any four numbers equidistant from the center is 130. The four corners and the four center numbers total 260. The sum of the four numbers forming a little square (2x2) is 130.

52	61	4	13	20	29	36	45
14	3	62	51	46	35	30	19
53	60	5	12	21	28	37	44
11	6	59	54	43	38	27	22
55	58	7	10	23	26	39	42
9	8	57	56	41	40	25	24
50	63	2	15	18	31	34	47
16	1	64	49	48	33	32	17

[1] See section on **Magic Squares.**

Irrational Numbers & the Pythagorean Theorem

Irrational numbers are numbers that cannot be expressed by an ending or a repeating decimal.

Examples are:

$\sqrt{2}$; $\sqrt{3}$; $\sqrt{5}$; π; $\sqrt{48}$; e; $\sqrt{235}$; φ ; . . .

When one attempts to write an irrational number as a decimal it turns out to be a never ending non-repeating decimal.

examples:

$\sqrt{2}$	$\approx 1.41421356..$
$\sqrt{235}$	$\approx 15.3297097...$
π	$\approx 3.141592653.$
e	$\approx 2.71828182..$
φ	$\approx 1.61803398...$ — the golden ratio

For thousands of years mathematicians have been devising methods to obtain more accurate decimal approximations of irrational numbers. With the use of high powered computers and infinite number series, their approximations can be carried out to any desired decimal accuracy. Considering the time and effort that has been expended in devising these methods, it is amazing that the _exact_ location for many irrational numbers can be found using the marvelous Pythagorean theorem. Ancient Greek mathematicians had proven the Pythagorean theorem[1], and had used it in constructing exact irrational number lengths.

---

[1]See section on **Pythagorean Theorem.** Note that π and e cannot be constructed using a straight edge and compass because besides being irrational numbers they are also transcendental numbers.

To locate the $\sqrt{2}, \sqrt{3}, \sqrt{4}, \sqrt{5}, \sqrt{6}, \sqrt{7}, \sqrt{8}, \ldots$ on the number line, construct right triangles with hypotenuses these lengths. Then use a compass to measure and to arc its location on the number line as illustrated.

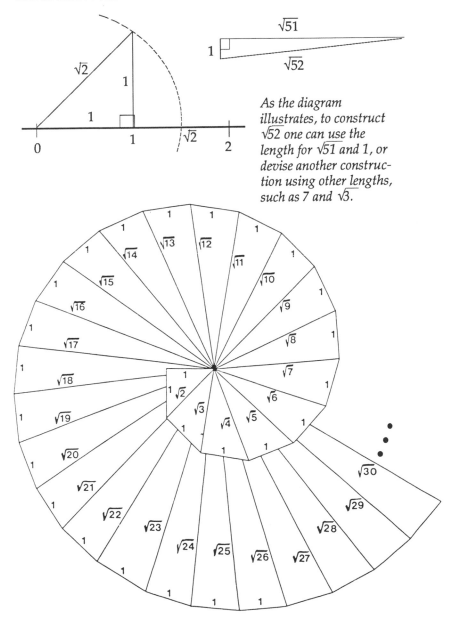

As the diagram illustrates, to construct $\sqrt{52}$ one can use the length for $\sqrt{51}$ and 1, or devise another construction using other lengths, such as 7 and $\sqrt{3}$.

Prime Numbers

A number is *prime* if it is a counting number greater than one whose only factors are itself and one.

```
                                                                    4486791661190433347949514103615951
778727200023729388613010364804475127856091580536317620183959201831046891496139730355336211345516747152878800071343434719468102573
2056939823423723521750452980127215084799977266875706902007262798468825185681532142985720637290299131726344453257416440944509835102
458816789016394945893695705164502436180232299551676032953391853644370456813506975908621980471189012253526094503315860264416805293
609502763225195411993787816118879503686070654670947060270393944504795918017973611326798274334403966485444764527043433252886593351
182627028129130176751736207356047170679601808698618919230547730826314303388930940240111607318421873986179317338167429362403908146
44653316799314240344162652439863548298861529638568560976139813700085452000021788527601812675169166661991965828892724054575480157275
83229363849690042394262629273440659534533488177062119409888429004222751020519807427651269590866648002084933159318354473037843867361
9005441075619665362094067700417583472704239335761771260615092329687186048219014455376174809648109097537444469206098114971853122623
214493115199270347679840290650446674341158632712607371820441076129040576076136413981094443341749359455758269556305513864442496417
3858932643106235024425974988499542755111318125373546749571089212977055534150530631766803652545011561999470470308097500183103909632
8170193148671666692203695843622915919803633402511820948390456453052580945876198697254235294719583716788165976228557375284260972377
34624468532362032697812178869122804644491770794318538447605771181055982286531205728018094382507023117903201429702131750687465985253
431776941579925010154034590557560961501317943693425717129107354716647460894659213261271724426089556159421057095612593905012539058
44971090566439919764443572232090088277547132202284397757577307755156586814029292609745277594756199575226844264383084190316769075728
9694016151292973116618877453197394261300350328032528435929861943078814647724393409604332129915703479716400017511529010473000220
9922649639362263334540500936114495516334353448137856451309799339035379297516197866955056300242370616123691536295773054211505
5910456824057243043107264563487603354791817176704413651372729793611485207957283097627765933169252913861287725833830000755907168
4963736432018866277260591886824978425490214803101062206217608651205629204397566858081153617726327091318013871586725598473961
011785486557173089001587860032432642701917412170832606266885767524270960067792953274073860017055574112385644597787359211013123
9116174462947424279402757853073650260508072041390081331521202378491760590327555321479818061901889766119669234837400805849571180
304127951763426872458894003176440733448134074320769587503577201360730815188490726917304830142391116460009314135465642597954514
070657714671682820664340999079994154553744192341743219322342556225337330000307729149794958036946844731378472355395291175411510838
529005489555874573161379011394384037750925611475414864709074417123434984529260791950700627431494379580386488091313784728539087088
0026316637158434572092613446646349461471539477576137139523005742738711369436082760696427026543527702736136135869093260269725551
3779540613157207839639234711716302148931471912295396724184965003568651689763453287082941395947177900964801256481815583067060479
102592456105207951836551169131910359549740985470715054586019569497463035136607427812945110179293744408550439525755985238064138
6024334590095603113812702733756964145723411334253659374534874239353454753541562848353277020568735893276116304379664911400779518
40194972590243413512340367603000111479282576710476304410428283662170092730348276006446663733212590650041724900734009461807180
521511830337701127276954129337130614839275753123345971272945709547194620649206864894706146638724446863374322129686904174973082
1988128828734262137746332765587654962249224479021366073985542526766237205174310004056032971936353519436617170570444606980090536230206
572184198365476245185953735703901570544032529590370380252845566751153611623327992220792074701117059219529570771087337331179143
31109235310571941636526000751811870409677546232442344250371711420947553711559044657975806459087061511314859335333809003952706994
49754706017899880089291245016257754132770060039748309843458690757070105112749413681511445021677564148263000093943008357622
41030373543510732971094700974352901173721376034036403003797326084264985090962924426514168406854873708069128489306440173627901793458120
608352937336031719528557879855681940089001530377603116837305843298426749218571088428975393086808894893532943760609014053918821
315225488711871355530334478179710328749361217101890072047791219983006575605905027543266901734740190518767938024093054447655515058
60615566182333568500576040635350569160784772184155359287162675479541813295910251752292898135383307406724693311157087329533181550172
571241383069650510594200787626692244790213667399852526766237205174310004560633679109763391936128063084100451577010609
6715175437874251157420312179302326059837561013691228792446847335696268932928659959506615591181514200293654788349653231870009324048
607367479721056014559625296609693253207520823569015306692406886855051744455937391510171859964015820917134315804816271224232024120
89910712295930383477654192505497173735916956929737564557551752743410701062065311357351941301151443000158730804485703355596137321380
0627488133203147212423024424900900090013117712094235287027904902558269414477316218172757645529425064199884354959614763200793817
28997204475976180032792520107827575456349470701701392574625552529800868158533926116252582419538490429368571714145913418234736446807428977
26494606148118080133362117215059685259419540161033389737332940002459193301875202828281866318718229305167905870144451339017941
20332813488432789225762118020445804015850910946414842840649640269394165004760210370817982850495571836159055709917919149733735278914364
020458426027261754757926341449756920202000097341235665651665152982919391434391222266146212420141533630378838636292115829347560
89663783785467893012638041042143785487394846469092341179949504338667818125954541347246524623195481401316071628427281713043422786918
5631200192333698966933544162693913110417302565016946627544558875644345191269279600693951809271906450264294092857410828353511882751
```

At 9 p.m. on October 30, 1978 the above number was discovered to be the largest known prime number as of that date. After 1800 hours of computer time, Laura Nickel and Curt Noll (Hayward, CA, high school students) found the prime number 2^{21701}-1. Continuing independently, Curt Noll discovered a larger prime number, 2^{23209}-1, a few months later. In May of 1979 Harry Nelson of the Livermore Lab discovered a prime nearly twice the size of Noll's, namely 2^{44497}-1.

Although computers have been programmed today to locate prime numbers, the Greek mathematician Eratosthenes (275-194 B.C.) invented this technique of a numerical sieve to discover the prime numbers smaller than some given number. The diagram has the prime numbers circled that are less than 100.

The Sieve of Eratosthenes

~~1~~	2	3	~~4~~	5	~~6~~	7	~~8~~	~~9~~	~~10~~
11	~~12~~	13	~~14~~	~~15~~	~~16~~	17	~~18~~	19	~~20~~
~~21~~	~~22~~	23	~~24~~	~~25~~	~~26~~	~~27~~	~~28~~	29	~~30~~
31	~~32~~	~~33~~	~~34~~	~~35~~	~~36~~	37	~~38~~	~~39~~	~~40~~
41	~~42~~	43	~~44~~	~~45~~	~~46~~	47	~~48~~	~~49~~	~~50~~
~~51~~	~~52~~	53	~~54~~	~~55~~	~~56~~	~~57~~	~~58~~	59	~~60~~
61	~~62~~	~~63~~	~~64~~	~~65~~	~~66~~	67	~~68~~	~~69~~	~~70~~
71	~~72~~	73	~~74~~	~~75~~	~~76~~	~~77~~	~~78~~	79	~~80~~
~~81~~	~~82~~	83	~~84~~	~~85~~	~~86~~	~~87~~	~~88~~	89	~~90~~
~~91~~	~~92~~	~~93~~	~~94~~	~~95~~	~~96~~	97	~~98~~	~~99~~	~~100~~

procedure:

1) 1 is crossed out since it is not classified as prime.

2) Circle 2, the smallest positive even prime. Now cross out every 2nd number, i.e. all multiples of 2.

3) Circle 3, the next prime. Now cross out every 3rd number, i.e. all multiples of 3. Some may already be crossed out since they are also multiples of 2.

4) Circle the next open number, namely 5. Now cross out every 5th number.

5) Continue the process until all numbers up through 100 are either circled or crossed out.

The Golden Rectangle

The golden rectangle is a very beautiful and exciting mathematical object, which extends beyond the mathematical realm. Found in art, architecture, nature, and even advertising, its popularity is not an accident. Psychological tests have shown the golden rectangle to be one of the rectangles most pleasing to the human eye.

Ancient Greek architects of the 5th century B.C. were aware of its harmonious influence. The Parthenon is an example of the early architectural use of the golden rectangle. The ancient Greeks had knowledge of the golden mean, how to construct it, how to approximate it, and how to use it to construct the golden rectangle. The *golden mean, ∅* (phi), was not co-incidentally the first three letters of *Phidias*, the famous Greek sculptor. Phidias was believed to have used the

The Partenon in Athens, Greece.

golden mean and the golden rectangle in his works. The society of Pythagoreans may have chosen the pentagram as a symbol of their order because of its relation to the golden mean.

Besides influencing architecture, the golden rectangle also appears in art. In the 1509 treatise *De Divina Proportione* by Luca Pacioli, Leonardo da Vinci illustrated the golden mean in the make up of the human body. The use of the golden mean in art has come to be labeled as the technique of *dynamic symmetry*.

Albrecht Dürer, George Seurat, Pietter Mondrian, Leonardo da Vinci, Salvador Dali, George Bellows all used the golden rectangle in some of their works to create dynamic symmetry.

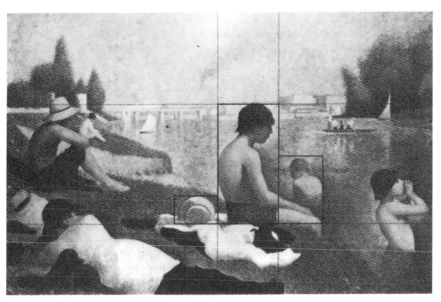

Bathers (1859-1891) by French impressionist George Seurat. There are three golden rectangles shown.

When the geometric mean is located on a given segment, AC, the golden mean[1] is formed,
so that

(|AC|/|AB|)=(|AB|/|BC|),
then |AB| is the *golden mean,*
also known as the *golden section,*
the *golden ratio,* or the *golden proportion.*

A _____ B _____ C

---

[1]To determine the value of the golden ratio, one must solve the equation $(1/x) = (x/(1-x))$, where x=|AB|, |AC|=1, and |BC|=(1-x). The golden ratio, |AC|/|AB| or |AB|/|BC| comes out to be — $[(1+\sqrt{5})/2] \approx 1.6$.

Once a segment has been divided into a golden mean, the golden rectangle can easily be constructed as follows:

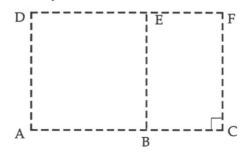

1) Given any segment $\overline{AC}$, with B dividing the segment into a golden mean, construct square ABED.

2) Construct $\overline{CF}$ perpendicular to $\overline{AC}$.

3) Extend ray $\overrightarrow{DE}$ so that line $\overleftrightarrow{DE}$ intersects line $\overleftrightarrow{CF}$ at point F. Then ADFC is a golden rectangle.

A golden rectangle can also be constructed without already having the golden mean, as follows:

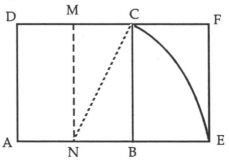

1) Construct any square, ABCD.

2) Bisect the square with segment $\overline{MN}$

3) Using a compass, make arc $\overset{\frown}{EC}$ using center N and radius |CN|.

4) Extend ray $\overrightarrow{AB}$ until it intersects the arc at point E.

5) Extend ray $\overrightarrow{DC}$.

6) Construct segment $\overline{EF}$ perpendicular to segment $\overline{AE}$, and ray $\overrightarrow{DC}$ intersects ray $\overrightarrow{EF}$ at point F. Then ADFE is a golden rectangle.

The golden rectangle is also *self-generating*. Starting with golden rectangle ABCD below, golden rectangle ECDF is easily made by drawing square ABEF. Then golden rectangle DGHF is easily formed by drawing square ECGH. This process can be continued indefinitely.

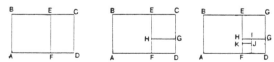

Using the final product of these infinitely many golden rectangles nestled in one another the *equiangular spiral* (also called the *logarithmic spiral*) can be made. Using a compass and the squares of these golden rectangles, make arcs which are quarter circles of these squares. These arcs form the equiangular spiral.

NOTE:

The golden rectangle continually generates other golden rectangles and thus outlines the equiangular spiral. The intersection of the diagonals pictured is the pole or center of the spiral.

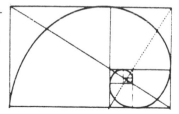

O is the center of the spiral.

A radius of the spiral is a segment with endpoints the center O and any point of the spiral.

Notice that each tangent to the point of the spiral forms an angle with that point's radius, e.g. T_1P_1O. The spiral is an equiangular spiral if all such angles are congruent.

This is also called a logarithmic spiral because it increases at a geometric rate, .i.e. a power of some number and a power or exponent is another name for logarithm.

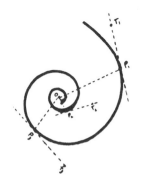

The equiangular spiral is the only type of spiral that does not alter its shape as it grows.

Nature has many forms of packaging — squares, hexagons, circles, triangles. The golden rectangle and the equiangular spiral are two of the most aesthetically pleasing forms. Evidence of the equiangular spiral and the golden rectangle are found in starfish, shells, ammonites, the chambered nautilus, seedhead arrangement, pine cones, pineapples, and even the shape of an egg.

Equally exciting is how the golden ratio is linked to the Fibonacci sequence. The limit of the sequence of ratios of consecutive terms of the Fibonacci sequence — (1,   1,   2, 3, 5, 8, 13, ... , $[F_{n-1}+F_{n-2}],...$) — is the golden mean, φ .

$$\frac{1}{1},\frac{2}{1},\frac{3}{2},\frac{5}{3},\frac{8}{5},\frac{13}{8},\frac{21}{13},\dots,\frac{F_{n+1}}{F_n} \to \varphi$$

$$1,\quad 2, 1.5, 1.6, 1.625, 1.6153, 1.619,\dots$$

$$\varphi = \frac{1+\sqrt{5}}{2} \approx 1.6$$

Besides appearing in art, architecture and nature, the golden rectangle is even used today in advertising and merchandising. Many containers are shaped as golden rectangles to possibly appeal to the public's aesthetic point of view. In fact, the standard credit card is nearly a golden rectangle.

Yet the golden rectangle interrelates with other mathematical ideas. Some of these are: infinite series, algebra, an inscribed regular decagon, Platonic solids, equiangular and logarithmic spirals, limits, the golden triangle, and the pentagram.

In a broad sense flexagons can be considered a type of topological puzzle. They are figures made from a sheet of paper, but end up having a varying number of faces which are brought to view by a series of flexings.

Making a Tri-Tetra Flexagon

The object below is a called a tri-tetra flexagon. **Tri** stands for the number of faces and **tetra** for the number of sides of the object.

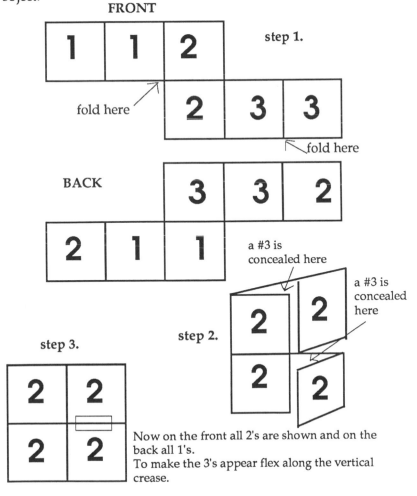

FRONT

step 1.

fold here

fold here

BACK

a #3 is concealed here

a #3 is concealed here

step 2.

step 3.

Now on the front all 2's are shown and on the back all 1's.
To make the 3's appear flex along the vertical crease.

Finding infinity in small places

Can you imagine what **infinity** is? Infinity is a never ending amount. The concept of infinity is hard to grasp. While we can easily grasp that the number 7 can describe 7 apples, and the number 1 billion (written 1,000,000,000) can describe the number of grains of sand in a jar. But an **infinite amount** is endless. A very exciting way to get a physical feeling of infinity is to hold up a mirror directly in front of another larger one. What happens is that you will see a mirror inside a mirror inside a mirror inside a mirror.... never ending.

Many people think that an infinite amount must take up a great deal of space, but on this small line segment AB, A_____B, there are an infinite number of points.

To prove this, we use the idea that between any two points another point can be found. So if point A and point B are on a line segment, then point C can be found between them. Then between pair A and C another point can be found, as well as between the pair C and B another point can be found. This process of finding another point between any two continues forever, so there are an infinite number of points on segment AB.

Another way to describe an infinite amount is by using the **flea story.**

*Half-flea wants to jump across the room. His friend tells him that he would never reach the other side if he promised to always jump in leaps that are only 1/2 the remaining distance. Half-flea says he would have no trouble reaching the other side. He first jumps 1/2 way across, then 1/2 way across the remaining distance, then 1/2 way across the next distance, and so on. Even though he is so very close to the other side he has to follow the **rule:** each jump must be 1/2 the remaining distance. Half- flea finally realizes that there will always be a remaining distance he would have to jump 1/2 way, and this would go on forever unless he gives up.*

Thus, even though **infinity** is an endless amount that cannot be identified by a number, we have discovered that it can fit in a very small space as well as a very large space.

The Five Platonic Solids

Platonic solids are convex solids whose edges form congruent regular plane polygons. *Only five such solids exist.*

The word solid means any 3-dimensional object, such as a rock, a bean, a sphere, a pyramid, a box, a cube. There is a very special group of solids called regular solids that were discovered in ancient times by the Greek philosopher, Plato. A solid is regular if each of its faces is the same size and shape. So a cube is a regular solid because all its faces are the same size squares, but this box, on the right, is not a regular solid because its faces are not all the

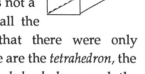

same size rectangles. Plato proved that there were only **five** possible regular convex solids. These are the *tetrahedron,* the *cube* or *hexahedron,* the *octahedron,* the *dodecahedron,* and the *icosahedron.*

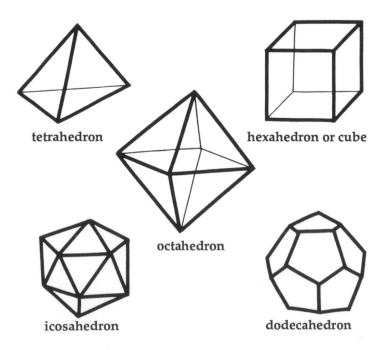

tetrahedron

hexahedron or cube

octahedron

icosahedron

dodecahedron

Here are patterns for making all five regular solids. Why not copy them, cut them out and try to fold them into their 3-dimensional forms?

tetrahedron

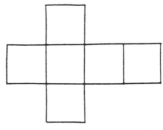

hexahedron or cube

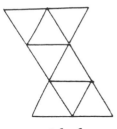

octahedron

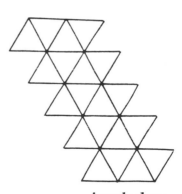

icosahedron

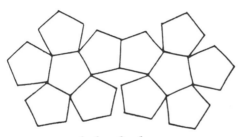

dodecahedron

The Pyramid Method for making magic squares

The pyramid method is one of the methods used for constructing odd order magic squares. The example below illustrates how to make a 5x5 magic square.

Procedure:
1) Sequentially place the numbers 1 to 25 of the magic square along diagonal boxes, as shown;
2) Relocate any number landing outside the magic square in an imaginary square in their respective positions in the magic square (hollowed numbers were relocated).

Although Plato is attributed with the discovery of the five Platonic solids (tetrahedron, hex-

The Kepler – Poinsot Solids

ahedron or the cube, octahedron, dodecahedron, icosahedron) and Archimedes is attributed with the Archimedean solids, these four non-convex solids were unknown to the ancient world. Kepler discovered his two in the early 1600's and Louis Poinsot (1777-1859) rediscovered these two and discovered two more in 1809. Their shapes are often used today as light fixtures and lamp shades.

small stellated dodecahedron

great stellated dodecahedron

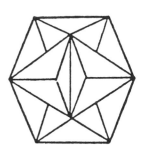

great dodecahedron

great icosahedron

The False Spiral Optical Illusion

This diagram appears to be a spiral, but closer examination shows it is composed of concentric circles. The *unit of direction* was discovered by Dr. James Fraser, and it was first described in the *British Journal of Psychology* in January 1908. It is also often called the *twisted cord* effect. Two cords of contrasting colors are twisted to form a single cord. It is then superimposed on different backgrounds. The illusion created is so convincing that even tracing the concentric circles to destroy the illusiion of a spiral or a helix is a difficult task.

The golden rectangle appears in so many facets of our lives — architecture, art, nature and sciences, as well as

The icosahedron & the golden rectangle

mathematics. Luca Paoioli's book, *De Divina Proportione* (illustrated by Leonardo da Vinci in 1509), presents fascinating examples of the golden mean in plane and solid geometry. The drawing below is such an example. Here, three golden rectangles intersect each other symmetrically and each perpendicularly to the other two. The corners of these rectangles coincide with the twelve corners of a regular icosahedron.

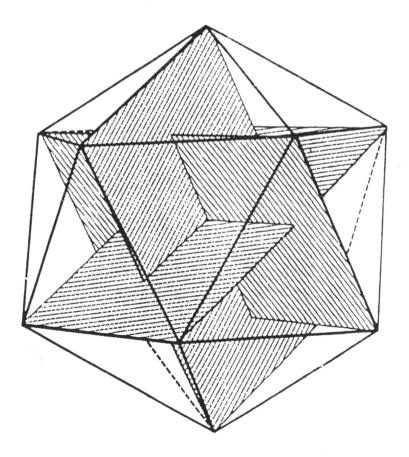

Zeno's Paradox — Achilles & the tortoise

Paradoxes are interesting, entertaining and a very important part of mathematics. They emphasize how important it is to state and prove ideas carefully so there are no loopholes. In mathematics, we try to make mathematical ideas cover as many facets as possible, i.e. we try to generalize a concept and thereby make it apply to more objects. It is important to generalize, but it can be dangerous. One must proceed cautiously. Some paradoxes illustrate this danger.

In the 5th century B.C., Zeno, using his knowledge of infinity, sequences and partial sums, developed this famous paradox. He proposed that in a race with Achilles, the tortoise be given a head start of 1000 meters. Assume Achilles could run 10 times faster than the tortoise. When the race started and Achilles had gone 1000 meters, the tortoise would still be 100 meters ahead. When Achilles had gone the next 100 meters the tortoise would be 10 meters ahead.

Zeno argued that Achilles would continually gain on the tortoise, but he would never reach him. Was his reasoning correct? If Achilles were to pass the tortoise, at what point of the race would it be?

See the appendix for the solution to *Achilles & the tortoise*.

Eublides' & Zeno's paradoxes
The Greek philosopher Eublides argued that one could never have a pile of sand. He proposed that certainly one grain of sand does not constitute a pile of sand. And if one adds another grain of sand to the one, they do not make a pile. He said if you do not have a pile of sand and if by adding one to what you have you still do not have a pile, then you will never have a pile of sand.

Along the same line of reasoning, Zeno looked at the points on a line segment. He argued if a point has no size, then adding another point to it will still have no size. Thus one will never get an object of any size by joining points. But he further argued that if a point does have size then a line segment must be infinitely long because it is composed of an infinite number of points.

The Mystic Hexagram

Mathematics is an unending treasure of intriguing ideas. This particular theorem was proven by the French mathematician Blaise Pascal (1623-1662) when he was sixteen years old. He named it the **mystic hexagram.**

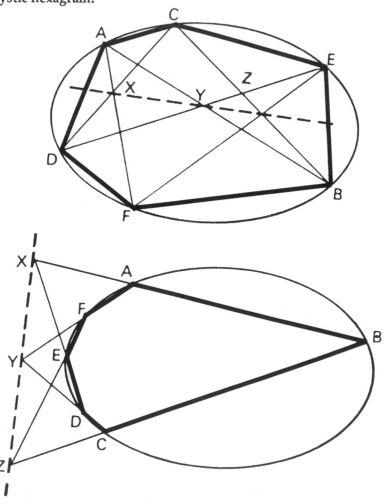

If a hexagon is inscribed in a conic, then the intersection of the three pairs of opposite sides are collinear.

Turn the triangle of pennies upside down by sliding one penny at a time to a new location so that it touches two other pennies.

The Penny Puzzle

The minimum number of moves needed is three.

Tessellations

Tessellations of a plane simply means being able to cover the plane with flat tiles so that no gaps and no tiles overlap. Given certain shapes one can use mathematics to decide ahead of time whether it would be possible without actually laying out the tiles. To figure this out one must know the mathematical fact that a circle has 360º.

Equipped with this tool and some geometry let's consider covering (tessellating) a floor with regular pentagons. A regular pentagon has five congruent sides and five congruent angles. To figure out the measure of the angles of a pentagon, divide the pentagon into triangles, as illustrated. For any triangle, its three angles total 180º. All of the five triangles composing the pentagon are congruent, since their corresponding sides and angles are congruent. We can now determine the measure of the pentagon's angles to be 108º. Thus when we try to put identical regular pentagons side by side we find we will have a gap because the pentagons cannot complete a circle or 360º (108º + 108º + 108º =324º).

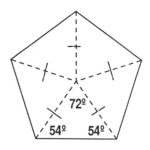

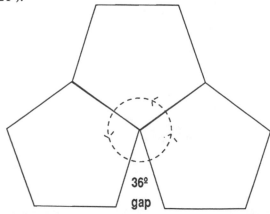

36º
gap

Now let's try to tessellate a floor using equilateral triangles. Each angle in a equilateral triangle is 60°. We see we can lay six identical equilateral triangles together, and they will complete the circle.

What about using squares, hexagons, octagons, or a combination? Here are some tessellations of planes.

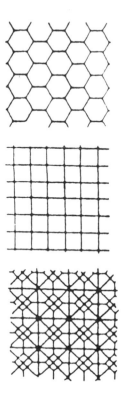

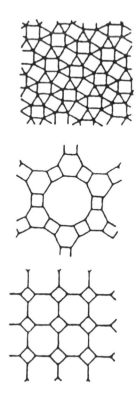

In a similar way, space can be tessellated – tiled with three-dimensional solids. Pictured below are truncated octahedra. They are the only Archimedean polyhedra that can be used to fill space while leaving no gaps and using no other shaped solids.

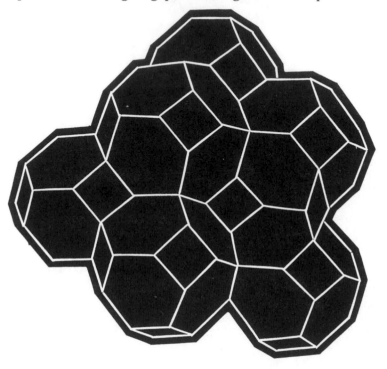

The renowned Dutch artist M.C. Escher used many mathematical concepts in his work — the Moebius strip, geodesics, projective geometry, optical illusions, the tribar, trefoil and tessellations — to name some. A number of his famous works use exciting tessellations which he created, for example, *Metamorphosis, Horseman, Smaller and Smaller, Square Limit, Circle Limit*. In addition to art, the study and applications of tessellating space is of special interest to the field of architecture, interior design, commercial packaging.

Diophantus is often called the father of algebra. Little is known about his life except that he lived between 100 to

Diophantus' Riddle

400 A.D. However, his age at death is known because one of his admirers described his life in an algebraic riddle.

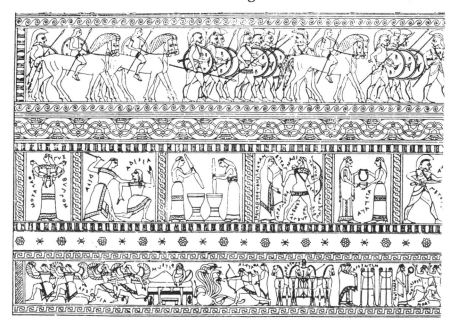

Diophantus' youth lasted 1/6 of his life. He grew a beard after 1/12 more of his life. After 1/7 more of his life, Diophantus married. Five years later he had a son. The son lived exactly 1/2 as long as his father, and Diophantus died just four years after his son's death. All of this totals the years Diophantus lived.

See appendix for solution to *Diophantus' Riddle*.

The Königsberg Bridge Problem & Topology

Topology originated with the solution in 1736 of a famous problem — *the Königsberg Bridge Problem.*

Königsberg[1] is a city on the Preger River that contains two islands and is joined by seven bridges. The river flows around the two islands of the town. The bridges run from the banks of

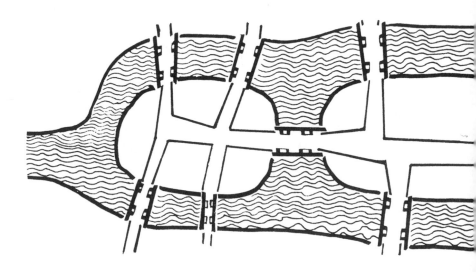

Diagram of the Königsberg Bridge Problem.

the river to the two islands in the river with a bridge connecting the islands. It became a town tradition to take a Sunday walk, and try to cross each of the seven bridges only once. No one had solved the problem until it came to the attention of the Swiss mathematician Leonhard Euler (1707-1783). At that time, Euler was serving the Russian empress Catherine the Great in St. Petersburg. In the process of solving this problem, Euler invented the branch of mathematics known as topology. He solved the *Königsberg Bridge Problem* by using an area of topology today called networks. Using networks, he proved that the problem of crossing each bridge of Königsberg only once was not possible.

This problem and Euler's solution launched the study of topology. Topology is a relatively new field. The mathematicians of the 19th century began delving into topology along with their studies of other non-Euclidean geometries. The first treatise of topology was written in 1847.

Euler.

---

[1] In the 18th century Konigsberg was a German city. Today it is Russian.

Networks

A *network* is basically a diagram of a problem. The network for the Königsberg Bridge Problem is illustrated below.

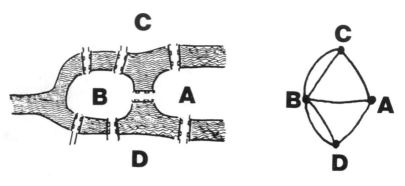

The network for the Könisgberg Bridge Problem.

A network consists of *vertices* and *arcs*. A network is *traveled* or *traced* by passing through all the arcs exactly once. A vertex may be crossed any number of times. The above diagram shows the vertices for the Königsberg Bridge Problem as A, B, C, D. Note the number of arcs passing through each vertex—A has 3, B has 5, C has 3, and D has 3. Since these are all odd numbers, these vertices are called *odd vertices*. An *even vertex* would have an even number of arcs passing through it. Euler discovered many properties about the number of odd and even vertices a network could have and still be traceable. Specifically, Euler noted that for a vertex to be odd, one would have to begin or end the journey at that vertex. With this in mind, he reasoned that since a network can have only one beginning and only one ending, that a traceable network could have only two odd vertices. Thus, since in the Königsberg Bridge Problem there are four odd vertices, it cannot be traced.

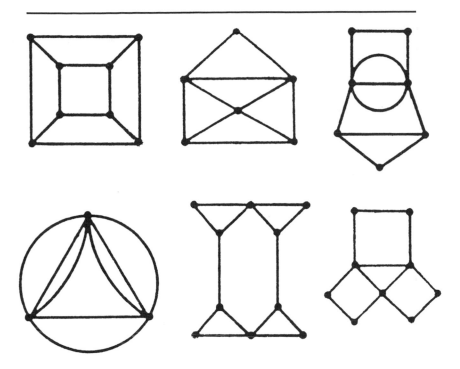

Which of the above networks is transversible (can be traced over without doubling back)?

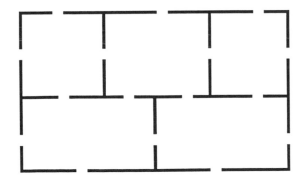

Can you find the path that passes through each door only once without lifting your pencil? Try to prove your solution by drawing a network.

Aztec Calendars

One of the earliest and most important calculating devices was the calendar — a system of measuring and recording the passage of time. Realizing nature furnishes a regular sequence of seasons which control food supplies, early people tried to correlate the solar day, the solar year and the lunar month. Since the lunar month is about 29.5 days while the solar year is 365 days 5 hours and 48 minutes and 46 seconds, it was not possible to get a whole number quantity of lunar months for a solar year. This is the major problem in developing consistent calendars. Even our present calendar is not consistent because every century year that is not divisible by 400 (e.g. 1700, 1800, 1900) must lose its extra leap day even though it is a leap year.

The Aztecs had two calendars — the religious calendar bore no relationship to the lunar month and the solar year. This calendar was important for ceremonial purposes, and Aztecs would include their birth dates as part of their names. The calendar consisted of twenty signs and thirteen numbers that followed in a fixed cycle of 260 days. Their second calendar was agriculturally oriented and contained 365 days.[1] Cyclical movements of heavenly bodies, enabled the Aztecs to regulate their calendar and accurately predict events such as eclipses.

---

[1]The Aztec borrowed many elements from the Toltec and Maya civilizations, including many parts of their calendar.

In 1790, the Aztec sunstone or stone calendar was discovered while repair work was being done on a cathedral in Mexico City. The cathedral had been built on the site of an ancient Tenochtitlàn pyramid-temple. The circular disc is 12 feet in diameter and weighs 26 tons. It recorded the history of the world according to the Aztec cosmology.

At the center, the sungod (Tonatiuh) is sculptured. Around the sungod the four suns or cosmogonic worlds (Tiger, Water, Wind and Rain of Fire) appear indicating the time before the Aztecs. Here the symbols of movement also appear. Another band of twenty glyphs — alligator, wind, house, lizard, serpent, death, deer, rabbit, water, dog, monkey, grass, reed, jaguar, eagle , vulture, earthquake, flint, rain, flower—represent the twenty days of the Aztec month.

The Impossible Trio

The beauty of a mathematical problem does not lie in its answer but in the methods of its solution. There exist problems in which the solution is finally determined to be no solution. Somehow *no solution* seems like a frustrating answer, but the thought process used to arrive at this conclusion is often most fascinating, and exciting discoveries of new ideas are made en route. So it was with the *three famous problems of antiquity.*

trisecting an angle —dividing an angle into three congruent angles

duplicating a cube — constructing a cube with twice the volume of a given cube

squaring a circle —constructing a square with the same area as a given circle

These problems stimulated mathematical thought and discoveries for over 2000 years until it was determined in the 19th century that the *three construction problems* were not possible using only a *straightedge and compass.* It was deduced that a straightedge can be used to construct line segments whose equations are linear (first degree equations), e.g. $y=3x-4$. A compass, on the other hand, can construct circles and arcs whose equations are second degree, e.g. $x^2 + y^2=25$. When these equations are solved simultaneously by using linear combinations they produce at most second degree equations. But the equations that are derived in solving the *three construction problems* by algebraic means are not first or second degree equations but rather cubic (a 3rd degree equation) or involve transcendental numbers. Therefore, a compass and a straightedge alone cannot be used to derive these types of equations or numbers.

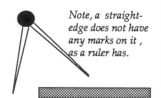

Note, a straightedge does not have any marks on it , as a ruler has.

Trisecting an angle

Certain specific angles such as 135º or 90º can be trisected using only a compass and straightedge. But given any angle, it is impossible to trisect it with only the use of a straightedge and compass because the equation used to solve this problem can be shown to be a cubic in the form $a^3 - 3a - 2b = 0$.

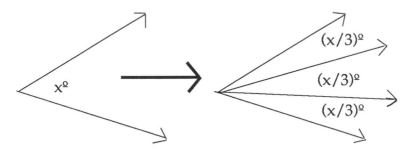

Duplicating a cube

In an attempt to duplicate a cube – that is double its volume – one might try to double the length of its sides. But this actually creates a cube whose volume is 8 times the given cube.

The volume of the cube
that is to be duplicated $= a^3$

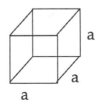

To duplicate this cube we want a cube whose volume is double or $2a^3$.

$$x^3 = 2a^3 \text{ which means } x = a\sqrt[3]{2}$$

Again we end up with a cubic we cannot construct with only a compass and straightedge.

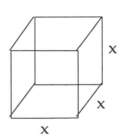

Squaring a circle

Given a circle with radius r, its area is πr^2.

So we want to construct a square whose area is πr^2.

$x^2 = \pi r^2$, which means $x = r\sqrt{\pi}$. Since π is a transcendental number it cannot be expressed in a finite number of rational operations and real roots, thus the circle cannot be squared using only a compass and straightedge.

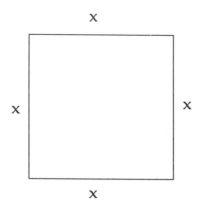

Although we see these three problems of antiquity are impossible to construct with only a compass and straightedge, ingenious methods and devises were created to solve them. And equally important, these problems stimulated the evolution of mathematical thought over the centuries. The conchoid of Nicomedes, the spiral of Archimedes, the quadratrix of Hippias, conic sections, cubic and quartic curves and several transcendental curves are some of the ideas which sprang from these three problems of antiquity.

Ancient Tibetan Magic Square

A 3 by 3 magic square is at the center of this ancient Tibetan seal, another example illustrating that mathematical ideas are not restricted to countries or by borders. The numerals appearing in the magic square are:

$$
\begin{array}{ccc}
4 & 9 & 2 \\
3 & 5 & 7 \\
8 & 1 & 6
\end{array}
$$

Perimeter, Area & the Infinite Series

The diagram below represents an infinite number of triangles. Each inscribed triangle is formed from the midpoints of the sides of the triangle that circumscribes it. To determine the sum of the perimeters of these triangles, we must look at the following series:

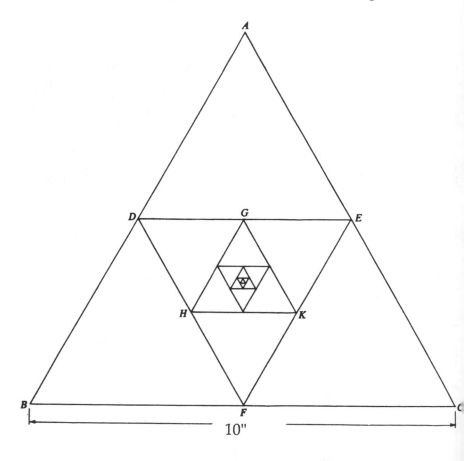

$$1/2 + 1/4 + 1/8 + 1/16 + 1/32 + 1/64 + 1/128 + ...$$

By looking at a number line one can determine the sum of these fractions.

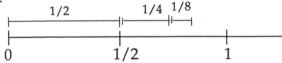

We notice that each successive fraction added to this series brings the sum closer and closer to 1, yet it will never go beyond 1. We conclude that this series has as its sum 1.

Now you may wonder how this information will help us determine the sum of the triangles' perimeters. First let's list the perimeters of each of the triangles:

$$30, 15, \ 15/2, \ 15/4, \ 15/8, \ 15/32, \ 15/64, \ 15/128,...\,[1]$$

Now we sum this series to determine the sum of the perimeters of the triangles.

$$30+15 +15/2 +15/4 +15/8 +15/16 +15/32 +15/64 +15/128 +...$$

simplifying the above we get

$$45 +15(1/2 + 1/4 + 1/8 + 1/16 +1/32 +1/64 + 1/128 +...)$$

now replacing 1 for the value of the sum of the series we get

$$45 +15(1) = 45 +15 = 60 \text{ as the perimeter.}$$

To determine the sum of the areas of the triangles is another challenge. You may have to do some research on the sum of a new infinite series.

---

[1]These values are determined using a theorem from geometry that states: *A segment joining the midpoints of two sides of a triangle is half the length of the side opposite it.*

The Checkerboard Problem

If two opposite corners of a checkerboard are removed, can the checkerboard be covered by dominoes?

Assume that each domino is the size of two adjacent squares of the checkerboard. The dominoes cannot be placed on top of each other and must lie flat.

See the appendix for the solution to the *checkerboard problem.*

Blaise Pascal (1623—1662) is known as one of France's famous mathematicians and scientists. He is credited with making many theo-

Pascal's Calculator

rectical mathematical and scientific discoveries, such as the theory of probability, the theory of liquids and hydraulic pressures. In addition, he invented this calculating machine at the age of eighteen. With the machine, one could add long columns of figures. Pascal's invention helped introduce fundamental principles that are the basis for modern calculators.

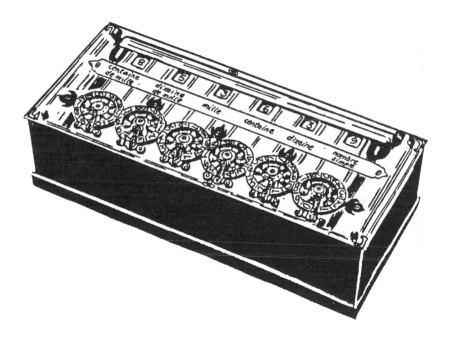

Isaac Newton & Calculus

Isaac Newton (1642-1727) was one of the inventors of calculus and of the theory of gravitation. Although Newton was a mathematical genius, he devoted much of his life to the study of theology. In 1665 the university he was attending at Cambridge was closed because of bubonic plague. He stayed home during this time, and developed his form of calculus, formulated the theory of gravitation and worked on other physical questions. Unfortunately, 39 years elapsed before his work was published.

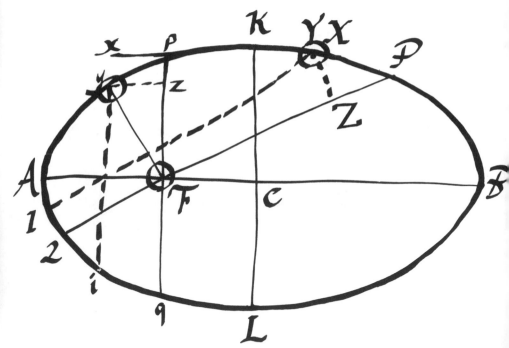

This is one of Newton's sketches showing the effects of gravity on an elliptical orbit.

It is important to always bear in mind that mathematics was developing simultaneously in various cultures in all parts of the world.

Japanese Calculus

For example, Seki Kowa, a 17th century Japanese mathematician, is credited with the development of Japanese calculus. It was called **yenri** (circle principle). The illustration below was drawn in 1670 by a pupil of Seki Kowa. It measures the circle's area by summing up a series of rectangles.

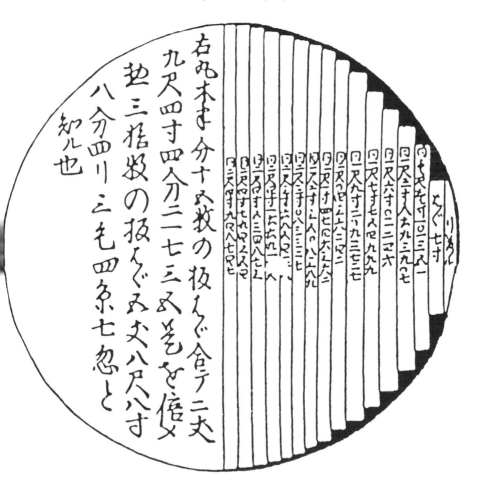

The proof of 1=2

The art of reasoning touches many aspects of our lives, be it deciding what to eat, using a map, buying a gift, or proving a geometric theorem. All sorts of skills and techniques enter into problem solving. A single flaw in our reasoning can create some rather strange or ridiculous results. For example, if you are a computer programmer, you dread overlooking a step that might lead to an infinite loop. Which of us has been certain of our explanation, solution or proof only to discover a mistake? In mathematics, *dividing by zero* is a common error which can cause outlandish results as illustrated by the proof of **1=2**. Can you find where the error is?

1=2 ?

If $a = b$ & $b, a > 0$, then $1 = 2$.

Proof

1) $a, b > 0$	*given*
2) $a = b$	*given*
3) $ab = b^2$	*x of $=$ step 2*
4) $ab - a^2 = b^2 - a^2$	*$-$ of $=$, step 3*
5) $a(b - a) = (b + a)(b - a)$	*factoring, step 4*
6) $a = (b + a)$	*$\div$ of $=$, step 5*
7) $a = a + a$	*substitution, steps 2 & 6*
8) $a = 2a$	*addition of like terms, step 7*
9) $1 = 2$	*$\div$ of $=$, step 8*

See the appendix for the flaw in *the proof of 1=2*.

The Symmetry of Crystals

Patterns and symmetry of natural phenomena abound. In 1912, physicist Max Von Laue passed X-rays through a spherical crystal onto a photographic plate. Dark points appeared that were arranged in perfect symmetry, which were later joined to form this design.

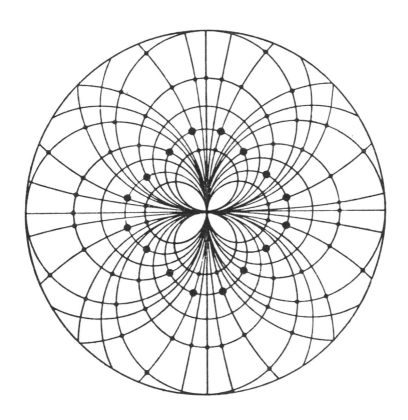

The Mathematics of Music

Music and mathematics have been linked from ancient times. During the medieval period the educational curriculum grouped arithmetic, geometry, astronomy and music together. Today's modern computers are perpetuating that tie.

Score writing is the first obvious area where mathematics reveals its influence on music. In the musical script, we find tempo (4:4 time, 3:4 time, and so forth), beats per measure, whole notes, half notes, quarter notes, eighth notes, sixteenth notes, and so on. Writing music to fit x-number of notes per measure resembles the process of finding a common denominator—the different length notes must be made to fit a particular measure at a certain tempo. And yet the composer creates music that fits so beautifully and effortlessly together in the rigid structure of the written score. When a completed work is analyzed, every measure has the prescribed number of beats using the various desired lengths of notes.

In addition to the obvious connection of mathematics to the musical score, music is linked to ratios, exponential curves, periodic functions and computer science. With ratios, the Pythagoreans (585-400 B.C.) were the first to associate music and mathematics. They discovered the connection between musical harmony and whole numbers by recognizing that the sound caused by a plucked string depended upon the length of the string. They also found that harmonious sounds were given off by equally taut strings whose lengths were in whole number ratios — in fact every harmonious combination of plucked strings could be expressed as a ratio of whole numbers. By increasing the length of the string of whole numbers ratios, an entire scale could be produced.   For example, starting with a string that produces the note C, then 16/15 of C's length gives B, 6/5 of B's gives A, 4/3 of A's gives G, 3/2 of G's gives F, 8/5 of F's gives E, 16/9 of E's gives D, and 2/1 of D's gives low C.

Have you every wondered why a grand piano is shaped the way it is? Actually there are many instruments whose shapes and structures are linked to various mathematical concepts.   Exponential functions and curves are such concepts. An exponential curve is one described by an equation of the form $y=k^x$, where $k>0$. An example is $y=2^x$. And its graph has the shape illustrated.

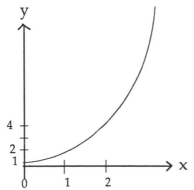

Musical instruments that are either string instruments or formed from columns of air, reflect the shape of an exponential curve in their structure.

The study of the nature of musical sounds reached its climax with the work of the 19th century mathematician John Fourier. He proved that all musical sounds — instrumental and vocal — could be described by mathematical expressions, which were the sums of simple periodic sine functions. Every sound has three qualities — pitch, loudness and quality — which distinguishes it from other musical sounds.

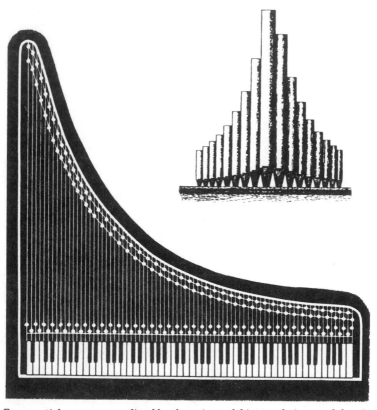

Exponential curves are outlined by the strings of this grand piano and the pipes of this organ.

Fourier's discovery makes it possible for these three properties of a sound to be graphically represented and distinct. Pitch is related to the frequency of the curve, loudness to the amplitude and quality to the shape of the periodic function.

Without an understanding of the mathematics of music, headway in using computers in musical composition and the design of instruments would not have been possible. Mathematical discoveries, namely periodic functions, were essential in the modern design of musical instruments and in the design of voice activated computers. Many instrument manufacturers compare the periodic sound graphs of their products to ideal graphs for these instruments. The fidelity of electronic musical reproduction is also closely tied to periodic graphs. Musicians and mathematicians will continue to play equally important roles in the production and reproduction of music.

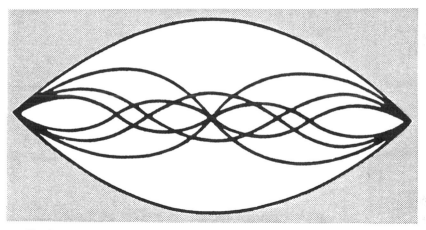

The diagram shows a string vibrating in sections and as a whole. The longest vibration determines the pitch and the smaller vibrations produce harmonics.

Numerical Palindromes

A *palindrome* is a word, verse, number, etc. that reads the same backwards as forward.

Examples are:
(1) *madam, I'm Adam*
(2) *dad*
(3) *10,233,201*
(4) *"Able was I ere I saw Elba"*

An interesting number curiosity is the following:

Start with any whole number. Add to it the number formed by reversing its digits. To this sum add the number formed by reversing the sum's digits. Continue this process, and see if eventually you will end up with a numerical palindrome.

Will you always end up with a numerical palindrome?

$$
\begin{array}{r}
1284 \\
+ 4821 \\
\hline
6105 \\
+ 5016 \\
\hline
11121 \\
+ 12111 \\
\hline
23232
\end{array}
$$

a numerical palindrome

A teacher announces that a test will be given on one of the five week days of next week, but tells the class, "You will not know which day it is until you are informed at 8a.m. of your 1p.m. test that day."

The Paradox of the Unexpected Exam

Why isn't the test going to be given?

See the appendix for the solution for *the unexpected exam paradox.*

Babylonian Cuneiform Text

The Babylonians probably adopted clay tablets and cuneiform (wedged-shaped) writing of the Mesopotamians because writing material, such as papyrus, was not available. Their number system was a base 60 positional number system which used the two symbols, Υ for 1 and $\blacktriangleleft$ for 10. $\blacktriangleright\!\!\blacktriangleleft$ = 60x10=600. Their clay tablet records showed evidence of sophisticated computation which they were able to carry out with their number system. This particular Babylonian problem and its solution were written during the reign of Hammurabi (1700's B.C.). The problem deals with lengths, widths and areas.

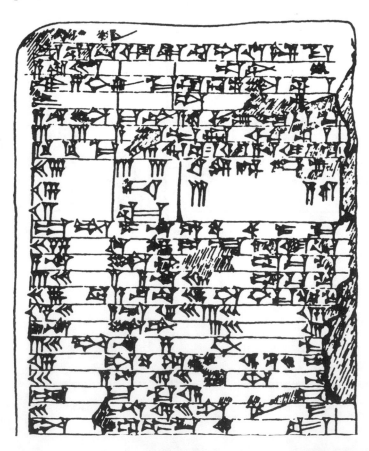

Spirals are forms which appear in many facets of nature, such as vines, shells, tornados, hurricanes, pine cones, the Milky Way, whirlpools.

The Spiral of Archimedes

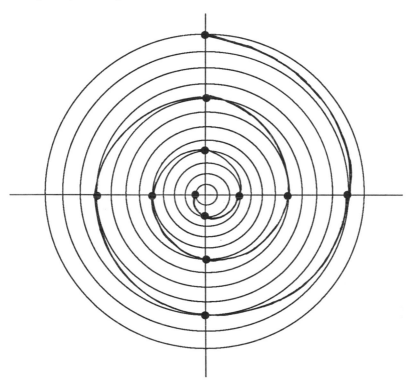

The Archimedean spiral is a two dimensional spiral. One way to envision it is to consider a bug crawling along a line which passes through the spiral's center, the pole. The bug crawls at a uniform rate while the line uniformly rotates about the pole. The path traced by the bug will be an Archimedean spiral.

Evolution of Mathematical Ideas

Pour les astres en général at pour les grand Comètes en particulier, trois mille ans ne sont pas grand'chose: dans le calendrier de l'éternité c'est moins qu'une seconde. Mais pour l'homme vous saves comme moi, mathematician lecteurer, que trois mille ans c'est boucoup beaucoup!

—Flammaron, 1892

It is easy to lose perspective of the fact that mathematics is an evolution of ideas beginning with the earliest discoveries given by prehistoric people dividing up their food and discovering the concept of number. Each contribution no matter how small is important to the development of mathematical thought. Some mathematicians have spent a lifetime on a single idea while others have varied their studies. For example, look at a condensed view of the development of Euclidean geometry. Geometric ideas were being discovered by a variety of peoples during ancient times. Thales (640 -546 B.C.) was considered the first to take a logical approach to geometric ideas. Others followed over a period of about 300 years, discovering much of the geometry studied in high school. In about 300 B.C., Euclid collected and organized the geometric ideas that had been created. It was an enormous task. He compiled all this information into a mathematical system which has come to be known as *Euclidean geometry.*. In his book, *The Elements,* he arranged the information so that it followed a logical progression. *The Elements,* written more than two thousand years ago, is far from a perfect mathematical system when scrutinized by today's mathematicians; but it remains a phenomenal piece of work.

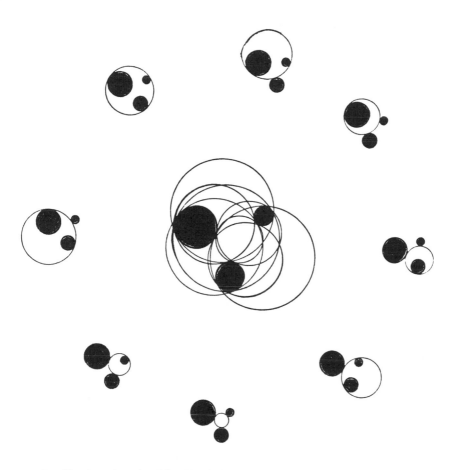

Apollonius, inspired by Euclid's work, made historical contributions to mathematics in the field of conics, astronomy and ballistics. The diagram above illustrates one of his intriguing problems:

Given three fixed circles,
find a circle tangent to all three.

The eight solutions are illustrated.

The Four Color Map Problem

Topology turns the tables on map coloring

It had always been an unproven rule among map makers that a map drawn on a flat surface or on a sphere required only four different colors to differentiate the various countries. In 1976, the *famous four color map problem* was supposedly set to rest with a computer "proof" by K. Appel and W. Haken of the University of Illinois, but their computer proof continues to be challenged. Interestingly, in 1974 W. McGregor, a graph theorist from New York, had made a map that supposedly cannot be colored with less than five colors. A direct challenge to the *four color map problem* conjecture.

The problem was to prove that every map drawn on a plane can be colored with only four colors so that adjacent territories have different colors.

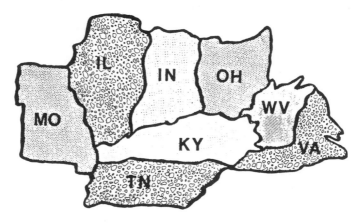

For an added twist, now consider map coloring on different topological models. Topologists study very unusually shaped surfaces — doughnuts, pretzels, Moebius shaped surfaces, and to them a sphere can be distorted into a plane by puncturing a hole in it and stretching and flattening it out. So in essence the number of colors required to color a plane and a sphere are the same. Topology is the study of properties of objects that remain unchanged when the objects are distorted — stretched or shrunk as on a rubber sheet. What type of properties remain unchanged under these conditions? *Since* distortion is allowed, we realize that topology cannot deal with size, shape or rigid objects. Some of the characteristics that topologists look for are the position or location of points inside or outside a curve, the number of surfaces of an object, whether an object is a simple closed curve, the number of inside and outside regions it has. So with these topological objects, **map coloring** is an entirely new problem since the four color map solution does not apply to them.

Try various map colorings on a strip of paper. Then twist it into a Moebius strip (give it a half-twist and join the ends together). Will four colors always be enough? Not likely.   What is the minimum number of colors

Moebius strip.

you can come up with that would work on any map. Try the same on a torus (which has a doughnut shape). The easiest way to experiment with this object is to envision the formation of a doughnut shape from a flat sheet of paper. Color a map on one side of the paper. Roll it into a cylinder. Then imagine bending the ends of the cylinder together so as to form a doughnut. Can you determine the minimum number of colors needed for a torus?

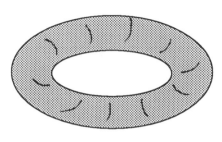

torus

Art & Dynamic Symmetry

There are many shapes in nature that are symmetrical — leaves, butterflies, the human body, snowflakes. Yet, there are many natural shapes that are not symmetrical. Consider the shape of an egg, of one wing of a butterfly, of the chambered nautilus, of the calico surfperch. These asymmetrical forms also possess a beautiful balance in their forms which has come to be

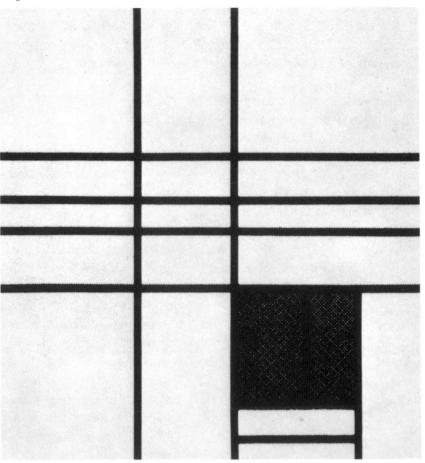

Composition with Yellow, 1936 by Mondrian. Mondrian is said to have approached every canvas in terms of the golden rectangle.

known as *dynamic symmetry.* The shape of the golden rectangle[1] or the proportion of the golden mean can be found in all shapes with dynamic symmetry.

The use of the golden mean and the golden rectangle in art is the *technique of dynamic symmetry.* Albrecht Dürer, George Seurat, Pietter Mondrian, Leonardo da Vinci, Salvador Dali, and George Bellows all used the golden rectangle in some of their works to create dynamic symmetry.

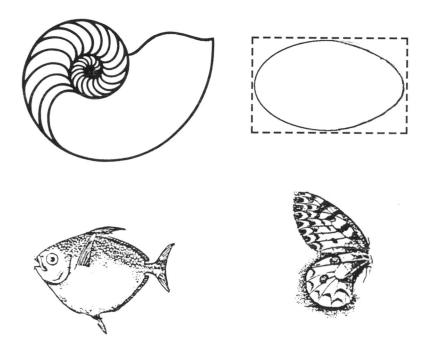

Illustrations show dynamic symmetry for the chambered nautilus, an egg, the wing of a butterfly, and the calico surfperch.

[1]See section **The Golden Rectangle.**

Transfinite Numbers

How many elements would you say are in the sets below:

$$\{a, b, c\} \ ?$$
$$\{-1, 5, 6, 4, 1/2\} \ ?$$
$$\{ \ \}$$

If you answered 3, 5, and 0 you are actually describing the cardinality of these sets.

Now, how many elements would you say are in the set,
$$\{1, 2, 3, 4, 5, \ldots\} \ ?$$

If you answered an infinite amount, you are not specific enough because there are various infinite sets. In fact, there is an infinite set of infinite cardinal numbers called transfinite numbers.

As the name implies, transfinite (to cross over the finite) cardinal is a "number" that describes an infinite amount. No finite number can adequately describe an infinite set. Two sets can be represented by the same cardinal number if the elements of one set can be paired off with the elements of the other set so no elements are left over in either set.

for example,

$\{a, b, c, d\}$
| | | | have cardinality 4, i.e. 4 elements in each set.
$\{1, 2, 3, 4\}$

set A=$\{1, 2, 3, 4, 5, \ldots, n \ldots\}$
set B=$\{1^2, 2^2, 3^2, 4^2, 5^2, \ldots, n^2, \ldots\}$

Set A and set B have the same cardinality because the elements of each set can be paired off as shown. Yet this seems paradoxical, since set A has numbers which are not perfect squares, but no elements are left out in the pairing off process.

19th century German mathematician George Cantor solved this paradox by creating a new number system – one that deals with infinite sets. He adopted the symbol, $\aleph$ (aleph – the first letter of the Hebrew alphabet), as the "number" of elements in infinite sets. Specifically, $\aleph_0$ (aleph null), is the smallest of transfinite cardinals.

$\aleph_0$ describes the number of elements in:

positive integers=$\{1, 2, 3, 4, 5, \ldots, n, \ldots\}$ n

whole numbers=$\{0, 1, 2, 3, 4, 5, \ldots, n-1, \ldots\}$ represents

positive integers= $\{+1, +2, +3, +4, +5, \ldots, n, \ldots\}$ a positive

negative integers= $\{-1, -2, -3, -4, -5, \ldots, -n, \ldots\}$ integer

integers=$\{\ldots, -3, -2, -1, 0, 1, 2, 3, \ldots\}$

rational numbers.

All of these sets and any others that can be paired off with the positive integers are considered to have cardinality $\aleph_0$

The examples below show methods for making one-to-one correspondences between the positive integers.

$\{1, 2, 3, 4, 5, \ldots, n, \ldots\}$ positive integers

$\{0, 1, 2, 3, 4, \ldots, n-1, \ldots\}$ whole numbers

$\{ 1, \quad 2, \quad 3, \quad 4, \quad 5, \quad 6, \quad 7, \quad 8, \quad 9, \ldots\}$ positive integers

$\{1/1, 2/1, 1/2, 1/3, 2/2, 3/1, 4/1, 3/2, 2/3, \ldots\}$

rational numbers

The chart below shows the order of how the rational numbers are placed in the preceding set.

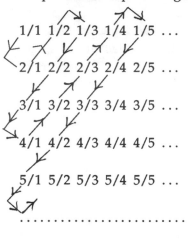

Cantor invented this method for arranging the rational numbers in some order so that every rational number would appear somewhere in this arrangement.

Cantor also developed an entire arithmetic system for operating with these transfinite numbers,

$$\aleph_0, \aleph_1, \aleph_2, \aleph_3, \ldots, \aleph_n, \ldots$$

He also proved that

$$\aleph_0 < \aleph_1 < \aleph_2 < \aleph_3 < \ldots < \aleph_n, \ldots$$

and that $\aleph_1$ describes the cardinality of real numbers, points on a line, points in a plane, and points in any portion of a higher dimension.

This logic problem dates back to eighth century writings.

Logic Problem

A farmer needs to take his goat, wolf and cabbage across the river. His boat can only accommodate him and either his goat , wolf, or cabbage. If he takes the wolf with him, the goat will eat the cabbage. If he takes the cabbage, the wolf will eat the goat. Only when the man is present are the cabbage and goat safe from their respective predators.

How does he get everything across the river?

See appendix for the solution of *the logic problem*.

The Snowflake Curve

The *snowflake curve* [1] derives its name from the snowflake-like shapes it assumes as it is generated. To generate a snowflake curve begin with an equi-

lateral triangle, figure 1. Then trisect each side of the triangle. Now on each of the middle thirds make an equilateral triangle pointing outward, but deleting the base of each new triangle that lies on the old triangle, figure 2. Continue this procedure with each equilateral triangular point — trisecting the sides and drawing new points, figure 3. Thus the snowflake curve is generated by repeating this process.

[1] See section on **Fractals** for additional information.

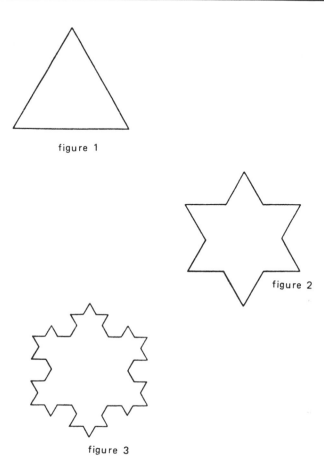

figure 1

figure 2

figure 3

*The astonishing characteristic of the snowflake curve is
that its area is finite while its perimeter is infinite.*

The perimeter is continually increasing without bound. This
curve can be drawn on a tiny piece of paper because its area is
finite, namely 1 3/5 times the area of the original triangle.

Zero— Where & When

The number **zero** is indispensable to our number system. But when number systems were being invented, they did not automatically include zero. In fact, the Egyptian number system did not have or require zero. Around 1700 B.C.

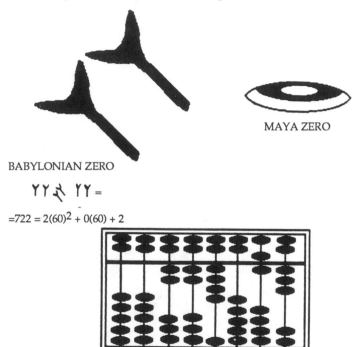

MAYA ZERO

BABYLONIAN ZERO

ᚖᚖ ᚔ ᚖᚖ =

=722 = 2(60)^2 + 0(60) + 2

ZERO ON THE ABACUS

the positional number system of base 60 evolved. The Babylonians used it in conjunction with their 360 day calendar. They performed sophisticated mathematical computations with it, but no symbol for zero was devised. An empty space was left in the number, standing for zero. About 300 B.C. Babylonians used this symbol for zero, ᚔ . The Maya and Hindu number systems evolved after the Babylonian. Their systems were the first to use a symbol for zero that functioned both as a place holder and as the number zero.

P_{appus' Theorem:} *If A, B, C, are points on line l₁ and D, E, F are points on line l₂; then P, Q, and R are collinear.*

Pappus' theorem & the nine coin puzzle

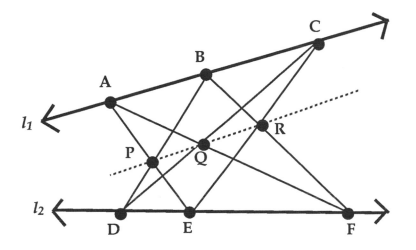

Apply Pappus' theorem to solve the *Nine Coin Puzzle.*

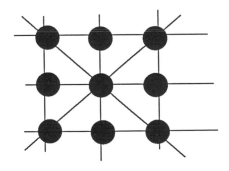

Nine Coin Puzzle: Rearrange these nine coins, which are 8 rows of 3, into 10 rows of 3.

See appendix for solution to *the nine coin puzzle.*

A Japanese Magic Circle

This Japanese magic circle is from the work of Seki Kowa. He was a 17th century Japanese mathematician who is credited with discovering a form of calculus and matrix operations used to solve systems of equations.

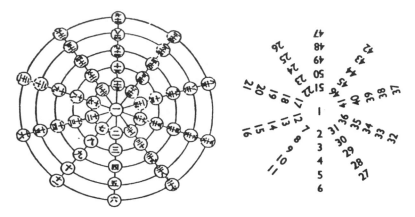

In the magic circle each diameter of numbers totals the same amount. The method used to form this magic circle seems similar to the method Gauss used to sum the first 100 natural numbers.

As the story goes, when Gauss was in primary school, his teacher gave the class the assignment to total the first 100 natural numbers. All the other students in the class busily began adding columns of numbers in the traditional manner of approaching this problem. Gauss sat at his desk thinking. Apparently his teacher thought he was daydreaming and asked him to get busy. Gauss replied that he had already solved the problem. The teacher asked for the solution, and Gauss illustrated his solution.

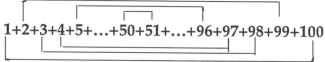

$$1+2+3+4+5+\ldots+50+51+\ldots+96+97+98+99+100$$

Gauss paired the numbers so the sum of each pair was 101, and he figured there were fifty pairs. Thus the sum came to 5050 = (50)(101).

There are a great many uses and applications of various geometric shapes to everyday needs and situations. One unusual example is

Spherical Dome & Water Distillation

on the island of Symi, off Greece, where a solar distillation unit in the shape of a hemisphere provides each of the island's 4,000 inhabitants with approximately one gallon of water per day.

The sun's heat evaporates water from the central supply of seawater. Then the fresh water condenses on the underside of a clear spherical dome, and drops of water slide down the sides and are collected along the edge of the dome.

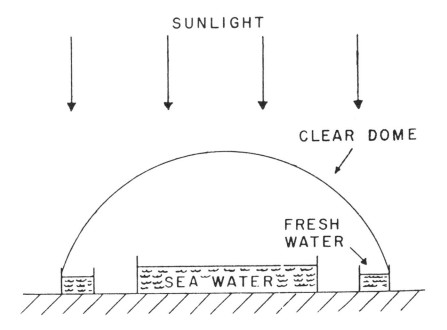

The Helix — Mathematics & Genetics

The helix is a fascinating mathematical object which touches many areas of our lives, such as genetic make-up, growth patterns, motion, the natural world and the manufactured world.

To understand the helix it is important to look at its formation. When a group of congruent rectangular blocks is joined lengthwise, a long rectangular column is formed. Perform the same process with rectangular blocks in which one face from each block is beveled. Then the resulting column curves around forming a circle. But if one face of each rectangular block were cut diagonally, the column would curve around on itself and form a 3-dimensional helix. Deoxyribonucleicacid, DNA—

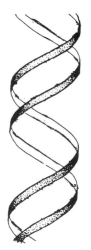
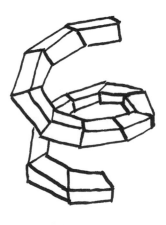

DNA double helix   Rectangular blocks cut
                          diagonally form a 3-D helix.

the inheritance chromosomes, are composed from two such 3-dimensional helices. DNA has two columns of sugar phosphate molecules that twist around joining skewed molecular units, as the above modified rectangular blocks.

There are different types of helices. In fact, the straight rectangular column and the circle column can be considered special cases of a helix. Helices can twist clockwise (right-handed) or counter-clockwise (left-handed). A clockwise helix, such as a corkscrew, when reflected in a mirror, appears to be a counterclockwise helix.

Examples of different types of helices appear in many facets of our world. Circular staircases, cables, screws, bolts, thermostat springs, nuts, ropes, and candy canes can be both right-handed and left-handed helices. Helices which spiral around cones are called conical helices, as seen in screws, bedsprings, and the spiral ramp designed by Frank Lloyd Wright in the Guggenhein Museum in New York.

formation of prochorite crystal

In nature we also find many forms of helices — the horns of antelopes, rams, narwhal whales and other mammals, viruses, some shells of snails and mollusks, the plant structure of stalks, of stems (e.g. peas), of flowers, of cones, of leaves and others. The human umbilical cord is a triple helix formed from one vein and two arteries that coil to the left.

It is not uncommon for right and left helices to intertwine. One botanical couple is the honeysuckle (left-handed) and the bindweed (right- handed which includes the morning glory in its family). They were immortalized by Shakespeare in *A Midsummer Night's Dream*. Queen Titania says to Bottom, "Sleep thou, and I will wind thee in my arm...So doth the woodbine (a common term for bindweed) the sweet honeysuckle gently twist."

Motion is yet another area where helices appear. Examples of helical paths are found in – tornados, whirlpools, draining water, a squirrel's path up or down a tree, and the clockwise spiral of the Mexican free-tailed bats of the Carlsbad caverns in New Mexico.

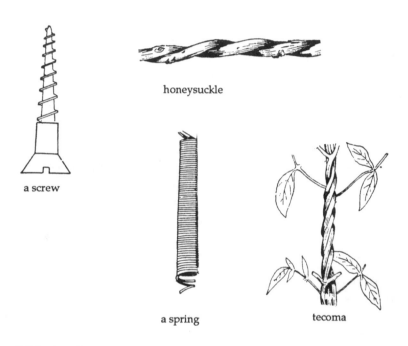

honeysuckle

a screw

a spring   tecoma

With the discovery that the helix is linked to the DNA molecule, it is not surprising to find the appearance of helices in so many areas. The varying forms of helices and their growth patterns in nature are themselves governed by a genetic code, and thus are continually generated by nature.

In the nineteen hundreds Claude F. Bradon discovered how magic squares could be

Magic "Line"

used to form artistically pleasing patterns. He discovered that if the numbers in a magic square were connected consecutively, they formed interesting patterns that have come to be known as *magic lines*. Actually a *magic line* is not a line, but a pattern that is formed. When these patterns are filled in with alternating shades, some very unique designs are created. As an architect, Bragdon used *magic lines* in architectural ornaments and graphic designs of books and textiles.

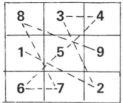

The magic line for lo shu, the earliest known magic square. It is from China in 2200 B.C.

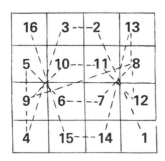

The magic line for Albrecht Dürer's magic square of 1514.

Mathematics & Architecture

We are all familiar with many of the mathematical forms used in architecture, such as the square, rectangle, pyramid and sphere. But there are some architectural structures which have been designed in less recognizable shapes. A striking example is the hyperbolic paraboloid used in the design of *St. Mary's Cathedral* in San Francisco. The Cathedral was designed by Paul A. Ryan and John Lee and engineering consultants Pier Luigi Nervi of Rome and Pietro Bellaschi of M.I.T.

St. Mary's Cathedral

At the unveiling when asked what Michelangelo would have thought of the Cathedral, Nervi replied: "He could not have thought of it. This design comes from geometric theories not then proven."

The top of the structure is a 2,135 cubic foot hyperbolic paraboloid cupola with walls rising 200 feet above the floor and supported by 4 massive concrete pylons which extend 94 feet into the ground. Each pylon carries a weight of nine million pounds. The walls are made from 1,680 prepounded concrete coffers involving 128 different sizes. The dimensions of the square foundation measure 255' by 255'.

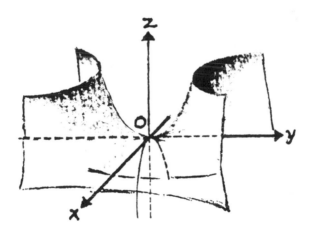

A hyperbolic paraboloid combines a paraboloid (a parabola revolved about its axis of symmetry) and a three dimensional hyperbola.

hyperbolic paraboloid equation:

$$\frac{y^2}{b^2} - \frac{x^2}{a^2} = \frac{z}{c} \qquad a,b>0, c\neq0$$

History of Optical Illusions

In the second half of the 19th century, there was a great surge of interest in the field of optical illusions. During this period nearly two hundred papers were written by physicists and psychologists describing optical illusions and why they occur.

Optical illusions are created by our eyes' structure, our mind, or a combination of both. What we see is not always what exists. It is important not to base conclusions strictly on what is perceived, but rather to verify by actual measurement.

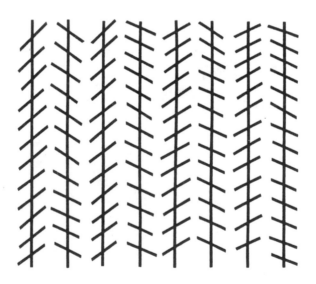

Zollner's illusion

It was this illusion that triggered the study of optical illusions in the 19th century. Johann Zollner (1834-1882), an astrophysicist and professor of astronomy (who made many contributions to the study of comets, the sun and the planets and was the inventor of the photometer) happened upon a piece of fabric with a design similar to the above drawing. The vertical lines are actually parallel, but certainly do not appear so. Some possible explanations for this optical illusion are:

1) *the difference between the acute angles which are set in different directions on the parallel segments.*
2) *the curvature of the eye's retina*
3) *superimposed segments cause our eyes to converge and diverge which makes the parallel segments curve.*

It was discovered that this illusion is most intense when the diagonal segments form angles of 45° with the parallel segments.

This famous optical illusion was created by cartoonist W.E. Hill and published in 1915. It is classified as an oscillation illusion because our eyes shift between two figures – an old woman and a young woman.

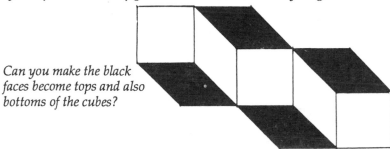

Can you make the black faces become tops and also bottoms of the cubes?

Trisecting & the equilateral triangle

Geometry has a wealth of ideas, concepts and theorems. It can be very interesting to discover a property which applies to a geometric object. For example, take *any triangle* and trisect its three angles. Then study the figure formed by the trisectors. What do you notice?[1]

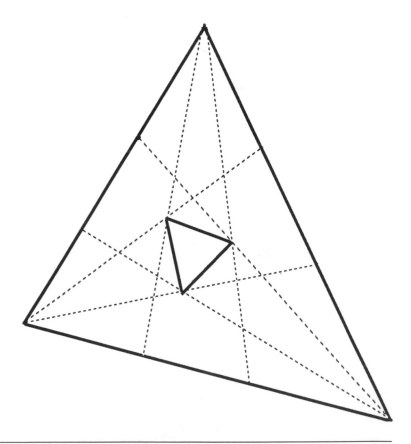

---

[1]It can be proven these trisectors *always form an equilateral triangle* regardless of the shape of the original triangle.

From each home three separate paths must be made – one to the well, one to the grain mill, and one to the woodshed. None of these paths cross each other. Can you solve this problem?

Wood, Water, Grain Problem

water

wood

grain

See appendix for the solution to *the Wood, Water, Grain Problem* .

Charles Babbage –the Leonardo da Vinci of modern computers

Considered the Leonardo da Vinci of modern computers, Charles Babbage (1792-1871) was an English mathematician, engineer, and inventor. Besides inventing the first speedometer, various precision machinery, codes and a method for identifying lighthouses from their beams, he spent most of

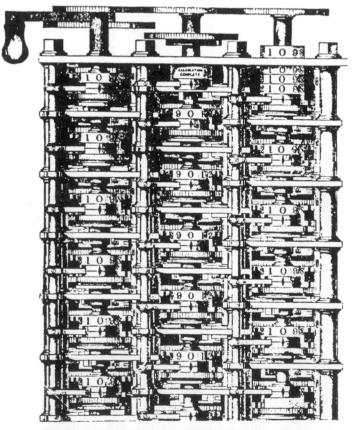

This illustration is a portion of Charles Babbage's difference engine which he started to build in 1823 and abandoned in 1842.

his time making a machine that would perform mathematical operations and calculate tables.

Babbage's original model of his *difference engine* was made with toothed wheels on shafts that were turned by a crank and could produce a table of squares up to 5 decimal places. Later Babbage designed a much grander machine with a 20 decimal place capacity that would stamp the answer on a copper engraver's plate. In the process of making parts, he became an expert technician, developed superior tools and techniques foreshadowing modern methods. He was continually perfecting parts and designs and scrapping old work. His perfectionism and the level of technology at that time kept him from completing the end product. When he abandoned the *difference engine*, he conceived the idea for the *analytical engine* [1] — capable of doing any mathematical operation, having a memory capacity of 1000 fifty digit numbers, using tables from its own library, comparing answers and making judgements not in the advance instructions. The machine's performance relied on mechanical parts and punched cards. Although his idea was never realized, the logical structure of his *analytical enginee* is used in today's computers.

The *analytical engine* actually represents a class of machines, in the same sense as today's *computer* is. It is quite astonishing Babbage singlemindedly pioneered this contemporary idea, engineered the machine, developed the tools to construct it, designed its various stages, and developed the mathematics needed to program it. A phenomenal task! A working model of the *analytical engine* was built by IBM as a tribute to Charles Babbage.

---

[1]Ada Lovelace, daughter of Lord Byron, encouraged and worked with Charles Babbage on his *analytical engine*. Besides monetarily contributing to his work, her keen knowledge of mathematics was invaluable in the computer programming aspect for the *analytical engine*. And equally important was her overall enthusiasm for his project.

Mathematics & Moslem Art

Since the representation of the human body was forbidden to Mohammedans, their art form was channeled into other areas. It was confined to ornament and mosaic, and concentrated on geometric designs. As a result, there is a definite connection between their art and mathematics.

The wealth of patterns that were created display:
- symmetries
- tessellations, reflections, rotations, translations of geometric forms
- congruences between dark and light patterns

This design shows translations of geometric forms.

This design illustrates tessellations[1] reflections, rotations, and symmetries. Congruences between light and dark forms are also highlighted by this design.

---

[1] A *tessellation* of a plane is the covering of the plane with a particular shaped tile so that no gaps are left and they do not overlap.

The Chinese magic square, illustrated below, is about 400 years old. In base ten numerals it translates to:

A Chinese Magic Square

27	29	2	4	13	36
9	11	20	22	31	18
32	25	7	3	21	23
14	16	34	30	12	5
28	6	15	17	26	19
1	24	33	35	8	10

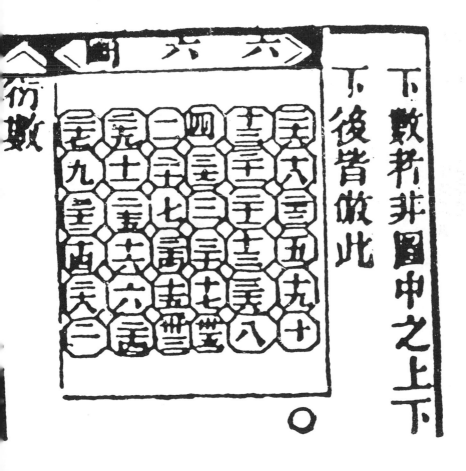

Infinity & Limits

The diagram below illustrates circumscribing regular polygons. The number of sides of the polygon successively increases. It would seem that the radii would grow without bound, but in fact the increasing radii approach a limit that is about 12 times that of the original circle.

There are ten stacks of ten silver dollars. You are given the weight of a real silver dollar, and are told each counterfeit coin

Counterfeit Coin Puzzle

weighs one gram more than a real silver dollar. You also know that one of the stacks is completely counterfeit, and you can use a scale that weighs by grams. What is the minimum number of weighings needed to determine the counterfeit stack?

See appendix for *counterfeit coin puzzle* solution.

The Parthenon –An Optical & Mathematical Design

Ancient Greek architects of the 5th century B.C. were masters in the use of optical illusions and the golden mean in the design of their buildings. These architects found that structures built precisely straight did not end up appearing straight to our eyes. This distortion is due to our retinas' curvature, causing straight lines that fall at particular angles to appear to curve when our eyes view them. The **Parthenon** is one of the most

The Parthenon

famous examples illustrating how these ancient architects compensated for the distortion caused by our eyes. As a result, the columns of the Parthenon actually curve outward, as do the sides of the rectangular base of the Parthenon. Diagram 1 illustrates how the Parthenon would have appeared if the architects had not made these adjustments.

Thus, by making these compensations, the structure and the columns appear straight and aesthetically pleasing.

diagram 1

Ancient Greek architects and artists also felt the golden mean and golden rectangle[1] enhanced the aesthetic appeal of structures and sculptures. They had knowledge of the golden mean — how to construct it, how to approximate it, and how to use it to construct the golden rectangle. The Parthenon illustrates the architectural use of the golden rectangle. Diagram 2 below shows how its dimensions fit almost exactly into the golden rectangle.

diagram 2

---

[1] See section on **Golden Rectangle** for additional information.

Probability and Pascal's Triangle

The triangle below formed from hexagonal blocks, has a unique way of generating the Pascal triangle. Balls from a reservoir on top pass down through

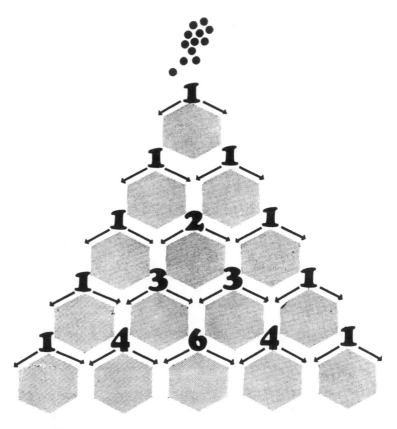

the hexagonal obstacles and collect below. At each hexagon there is an equal chance for a ball to roll to the right or to the left. As illustrated the balls distribute themselves according to the numbers of the Pascal triangle. The balls collect at the bottom and produce the bell shaped *normal distribution curve*. This curve is used — by insurance companies to set rates, in science to study the behavior of molecules, and in the study of population distribution.

Pierre Simon Laplace (1749-1827) defined the probability of an event to be the ratio of the number of ways in which the event can happen to the total possible number of events. Therefore when flipping a coin, the probability of getting a head is

$\dfrac{1}{2}$ – number of heads on a coin

$\phantom{\dfrac{1}{2}}$ – number of possible events

(head and tail)

The Pascal triangle can be used to calculate different combinations and the total possible combinations. For example, when four coins are tossed in the air the possible combinations of heads and tails are:

4 heads – HHHH=**1**
3 heads & 1 tail – HHHT, HHTH, HTHH, THHH = **4**
2 heads & 2 tails –HHTT, HTHT, THHT, HTTH, THTH,
                                          TTHH = **6**
1 head & 3 tails –HTTT, THTT, TTHT, TTTH = **4**
4 tails – TTTT =**1**

The fourth row from the top of the Pascal triangle indicates these possible outcomes — 1   4   6   4   1. The sum of these numbers represents the total possible outcomes = 1+4+6+4+1=16. Thus the probability of tossing 3 heads and 1 tail is :

$\dfrac{4}{16}$ – possible combination of 3 heads and 1 tail is

$\phantom{\dfrac{4}{16}}$ – total possible combinations.

For large combinations when the Pascal triangle would be tedious to extend, Newton's binomial formula can be used. Each row of the Pascal triangle contains the coefficients in the expansion of the binomial $(a+b)^n$. For example, to find the coefficients for $(a+b)^3$, one looks at the 3rd row from the top row (which is considered the zero row, i.e. $(a+b)^0=1$). At row three, one finds 1 3 3 1, which are the resulting coefficients,

$$1a^3 + 3a^2b + 3ab^2 + 1b^3 = (a+b)^3.$$

The binomial formula can be used for the nth row of the Pascal triangle.

binomial formula:

$$(a + b)^n = a^n + \frac{n}{1}(a^{n-1}b^1) + \frac{n}{1}\left(\frac{n-1}{2}\right)(a^{n-2}b^2) + ... + b^n$$

the rth coefficient   is $\dfrac{n!}{r!\,(n - r)!}$

the number of combinations of n objects taken r at a time is

$$C(n,r) = \frac{n!}{r!(n-r)!}$$

10 objects taken 3 at a time would be

$$C(10,3) \;=\; \frac{10!}{3!(10\text{-}3)!} \;=\; \frac{10\ 9\ 8\ 7\ 6\ 5\ 4\ 3\ 2\ 1}{3\ 2\ 1\ 7\ 6\ 5\ 4\ 3\ 2\ 1} = 120$$

120 possible combinations for 10 objects taken 3 at a time, or check the 10th row of the Pascal triangle.

As a rope is wound or unwound around another curve (here a circle), it describes an involute curve.

The Involute

Many examples of involutes appear in nature as found on the tips of a hanging palm leaf, the beak of an eagle, the dorsal fin of a shark.

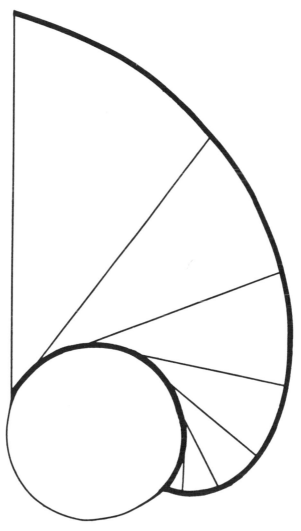

The pentagon, the pentagram & the golden triangle

Starting with a regular pentagon, the pentagram is generated by drawing in the diagonals of the pentagon. Within the shape of the pentagram exist golden triangles. These golden triangles divide the sides of the pentagram into golden ratios.

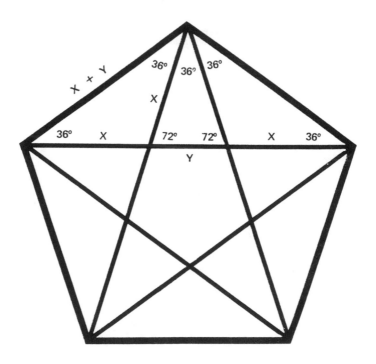

The golden triangle is an isosceles triangle with a vertex angle 36° and base angles 72° each. Its legs are in a golden ratio to its base. When a base angle is bisected, the angle bisector divides the opposite side in a golden ratio[1] and forms two smaller isosceles triangles.

One of these triangles is similar to the original triangle. The other triangle serves as the generator of the spiral's curve.

Continuing the process of bisecting a base angle of the new golden triangle generates a series of golden triangles and the formation of an equiangular spiral.[2]

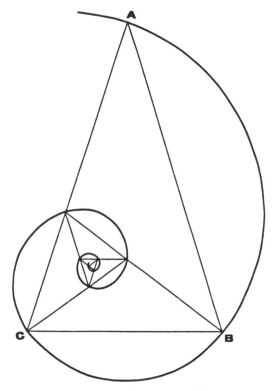

$$\frac{|AB|}{|BC|} = \text{golden ratio, } \phi = \frac{\left(1 + \sqrt{5}\right)}{2} \approx 1.6180339...$$

---

[1] See footnote on page 32 for additional information on the golden ratio.

[2] See **The Golden Rectangle** for information on the equiangular spiral.

Three men facing the wall

Three men are placed in a line perpendicular to a wall. They are blindfolded. Then three hats are taken from a bin of three tan hats and 2 black hats. The men are given that information. Then the blindfolds are removed. Each man is asked to determine what color hat he is wearing. The man

farthest from the wall who sees the two men and their hats in front of him says, *"I do not know which color hat I am wearing."* The second man from the wall who heard the reply and sees the man and the hat ahead of him says the same thing. The third man who sees only the wall, but has heard the two replies says,

"I know which color hat I am wearing."

Which color is he wearing and how did he determine it?

See the appendix for *The Three Men Facing the Wall* **solution.**

If a square is formed, using the sum of any two consecutive Fibonacci numbers as the length of its sides, an interesting geometric fallacy occurs.

Geometric fallacy & the Fibonacci sequence

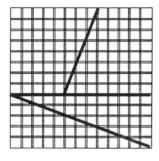
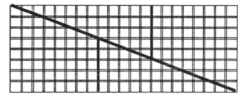

example:

1) Use the consecutive Fibonacci numbers 5 and 8.

2) Form a 13x13 square.

3) Cut it up as illustrated. Now calculate the square's area and the rectangle's area. The square's area is 1 unit greater than that of the rectangle.

4) Do the same process for the Fibonacci numbers 21 and 34. In this case, the area of the rectangle is 1 unit greater than the square.

The 1 unit discrepancy will alternate between the square and the rectangle, depending on which consecutive Fibonacci numbers are used.[1]

---

[1]The sequence of ratios formed from consecutive Fibonacci numbers 1/1, 2/1, 3/2, 5/3, 8/5, 13/8, ...F_{n+1}/F_n alternates in value above and below the golden ratio. The limit of this sequence is the value of the golden ratio $(1+\sqrt{5})/2$. For additional information see the section on **The Golden Rectangle.**

Mazes

Today we think of a maze as a recreational puzzle, but in earlier times mazes conjured up mystery, danger and confusion. One could surely get lost in the intricately winding paths or perhaps encounter monsters harbored in the maze's interior. In ancient times, labyrinths were often constructed to defend fortresses. Invaders would be forced to travel a great distance in a maze, and thus be exposed to attack.

Mazes have appeared in various parts of the world over the centuries.

- *stone carvings in Rock Valley, Ireland – circa 2000 B.C.*
- *Minoan labyrinth in Crete – circa 1600 B.C.*
- *Italian Alps, Pompeii, Scandinavia*
- *turf mazes of Wales and England*
- *mosaic mazes set in floors of European churches*
- *African textile mazes*
- *Hopi Indian rock carvings of Arizona*

Today, mazes are of special interest in the field of psychology and computer design. For decades psychologists have used mazes to study the learning behavior of animals and humans. Computerized robots have been designed to problem solve mazes as an initial step in the design of more advanced learning machines.

The garden at Hampton Court.

Topology is the field of mathematics that studies mazes as a branch of networks (solving problems by a method of diagramming). A Jordan curve is often mistaken for a maze. In topology we learn that a Jordan curve is a circle that has been twisted or bent inward and around without crossing itself. It has an inside and outside as does a circle, unlike a maze. Thus, the only way of getting from the inside to the outside is to cross the curve.

Since robots are used to solve mazes, systematic methods had to be devised to solve mazes.

methods for solving mazes

(1) For a single maze, shade blind alleys and loops you see. The remaining routes will be to the goal. Then select the most direct path. If the maze is complicated this method would be hard to apply.

(2) Go through the maze always keeping one hand (either the right or the left) in contact with the wall. This method is easy, but does not work for all mazes. Exceptions are (a) those mazes with two entrances and a path connecting them that does not pass through the goal, (b) mazes with paths that loop or surround the goal.

(3) French mathematician M. Trémaux has devised a general method to solve any maze.

procedure:

(a) As you go through the maze, constantly draw a line to the right.

(b) Whenever you come to a new juncture take any path you like.

(c) If on your new path, you come to an old juncture or dead end, turn and go back the way you came.

(d) If returning on an old path you come to an old juncture, take

This maze appears on a Navajo blanket.

any new path, if there is one. Otherwise take an old path.

(e) Never enter a path marked on both sides.

This method is supposed to be fool proof – but may take some time.

Whether with a life size scale or with a pencil in hand, mazes continue to challenge and intrigue the mind and provide enjoyment.

WATERLOO ROAD

This maze of London appeared in *The Strand Magazine* in April 1908 with the following instructions: *"The traveller is supposed to enter the Waterloo Road, and his object is to reach St. Paul's Cathedral without passing any of the barriers which are placed across those streets supposed to be under repair."*

In this illustration a checker board pattern of the Chinese counting

Chinese "Checkerboards"

board is shown. The Chinese were the first to evolve a system of set rules for solving systems of simultaneous equations.

They would place components in a checkerboard lay out, and then apply the rules which were based on matrices to solve the problem.

Conic Sections

It is often baffling to some people that mathematicians will pursue a problem or an idea simply because it is interesting or curious. Looking back at the ancient Greek thinkers, we find them studying ideas intensely without regard to their immediate usefulness, but because they were exciting, challenging or interesting. And so it was with their study of *conic sections.*

Their primary interest in these curves was to use them to help solve the three ancient construction problems — squaring a circle, doubling a square, and trisecting an angle. These problems had no practical value at the time, but they were challenging and stimulated mathematical thought. More often than not the practical use of a particular mathematical idea does not manifest itself for years. Conic sections created during the third century B.C., gave 17th century mathematicians foundations to begin to formulate various theories involving curves. For example, Kepler used ellipses to describe the paths of planets, and Galileo found the parabola to fit the motions of trajectiles on the earth.

The diagram on page 197 illustrates how a plane intersecting the double cone produces the *circle, ellipse, parabola,* and *hyperbola.*

question: How should a plane intersect the cone to produce a straight line, two intersecting lines, or a point?

There are many examples in the universe that form these curves. One contemporary and exciting example is **Halley's Comet.**

In 1704 Edmund Halley worked on the orbits of various comets for which data was available. He concluded that the comets of 1682, 1607, 1531, 1456 were a single comet orbiting the sun elliptically about every 76 years. He successfully predicted its return in 1758, and as a result it became known as Halley's Comet. Recent research suggests that Halley's Comet may have been recorded by the Chinese as early as 240 B.C.

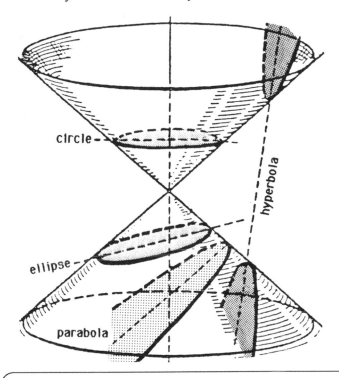

examples of these curves in the universe

parabola—
- arc of spouting water
- shape of flashlight's light on a flat surface

ellipse—
- orbital paths of some planets and some comets

hyperbola—
- paths of some comets and other astronomical objects

circle—
- ripples on a pond
- circular orbits
- the wheel
- objects in nature

The Screw of Archimedes

The Archimedean screw, when submerged in water and rotated, was able to pump water uphill.

It is still used for irrigation in various parts of the world.

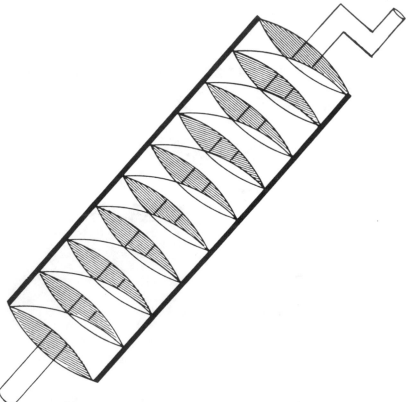

Archimedes (287-212 B.C.) was a Greek mathematician and inventor. He discovered the laws of lever and pulley. His discoveries led to the invention of machines capable of moving heavy loads easily. He is also credited with discovering a method to compare volumes by submerging objects in water — hydrostatics — buoyancy — the use of calculus ideas — inventing the catapult — inventing contoured mirrors to concentrate the sun's rays.

Optical illusions are created by our minds, our eyes' structure or both. For example, when we view a region that has both light and dark objects, the fluids in our eyes are not perfect-

Irradiation Optical illusion

ly clear and light scatters while passing to the retina at the rear of the eye (this is where the eye detects light). As a result, bright light, or light areas spill over onto dark areas of the image that is on the retina. Thus a light region will appear larger than a dark one of equal size, as in the diagram below. This explains why dark clothing, especially black, makes you appear more slender than if you were wearing light, or white clothing of the same design. This illusion is called *irradiation* , and was discovered by Herman L. F. von Helmholtz in the 19th century.

The Pythagorean Theorem and President Garfield

President James Abram Garfield (1831-1881), the twentieth president of the United States, was quite interested in mathematics. In 1876, while serving as a member of the House of Representatives, he discovered an interesting proof of the Pythagorean theorem[1]. The *New England Journal of Education* published the proof.

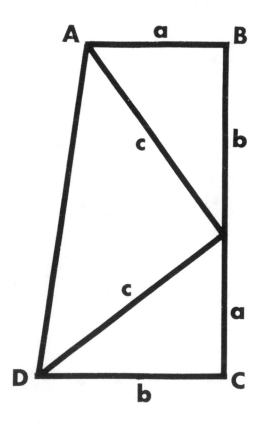

---

[1]See section on **Pythagorean Theorem.**

The proof uses two methods to calculate the area of a trapezoid.

method (1): area of a trapezoid = 1/2(sum of the bases)(altitude)

method (2): dividing trapezoid into 3 right triangle and calculating the area of these 3 right triangles.

proof

Construct trapezoid ABCD with $\overline{AB} \mid \mid \overline{DC}$, angles C and B right angles, and the indicated lengths, a, b and c.

Calculate the area of the trapezoid using the two methods listed above.

area method (1)	=	area method (2)
$1/2(a+b)(a+b)$		$=1/2(ab)+1/2(ab)+1/2(c\,c)$
$(a+b)\,(a+b)$	=	$ab + ab + c^2$
$a^2 + 2ab + b^2$		$= 2ab + c^2$
QED $a^2 + b^2$		$= c^2$

The Wheel Paradox of Aristotle

Two concentric circles are shown on this wheel. The wheel moves from A to B as it rotates once. Notice |AB| corresponds to the circumference of the large circle. Since the small circle also rotated once and traveled the distance |AB|, is not its circumference |AB| ?

Galileo's explanation of Aristotle's wheel paradox

Galileo analyzed the problem by considering two concentric squares on a square "wheel". As the box flips 4 times, (transversing the perimeter of the square wheel, |AB|), we notice the small box is carried along 3 jumps. This illustrates how the small circle is also carried along the distance |AB|, and that |AB| does not represent its circumference.

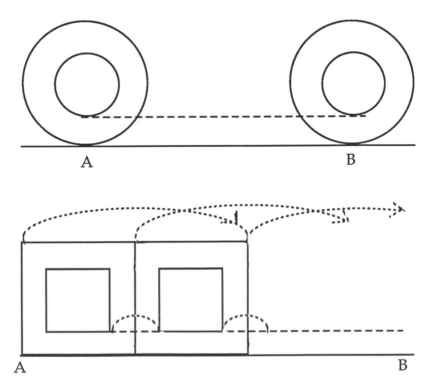

A   B

A   B

O n the Salisbury Plain in
England stands the awesome
stone structure known as

Stonehenge

Stonehenge. It was started around 2700 B.C., and the last of the
three stages was completed about 2000 B.C.

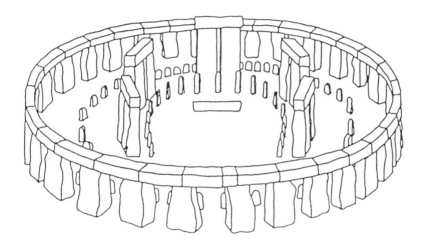

What was the purpose of Stonehenge? What meaning did it
have for the different groups of people who used and developed
it? Was it:

- a religious temple?
- a lunar and solar observatory for the
  winter solstice sunset and the summer
  solstice sunrise?
- a lunar calendar?
- a primitive computer for predicting lunar
  and solar eclipses?

Since there is no written documentation by the builders or users
of Stonehenge, there is no way to know its true purpose. From
the shreds of evidence, all theories remain speculation. What is
evident though, is that the builders had knowledge of a form of
measurement and geometry.

How many dimensions are there?

Art as varied as the early cave drawings, the Byzantine icon, Renaissance painters, and the Impressionists depict subjects that exist either in second or third dimension. Artists, scientists, mathematicians, and architects have developed their own renditions of how they believe certain objects would appear in the 4th-dimension. One example is this 4th-dimensional drawing of a cube, called a *hypercube*, by architect Claude Bragdon in 1913. Bragdon incorporated his hypercube drawing and other fourth dimensional designs into his works. An example is his Rochester Chamber of Commerce Building.

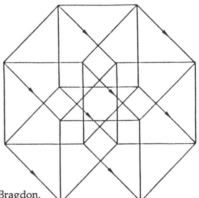

The hypercube by Claude Bragdon.

The existence of other dimensions beyond the third dimension has always been an intriguing notion. From a mathematical point of view, the likelihood of higher dimensions follows a logical progression of thought.

For example, begin with a *zero-dimensional* object, a *point*. Now move the point one unit to either the right or the left, and a line segment is formed. The *line segment* is a *one-dimenional* object.

Move the line segment one unit upward or downward, and a square is formed. The *square* is a *two-dimensional* object. Proceeding in the same manner, take the square and move it one unit outward or inward, and a *cube* is formed, which is a *three-dimensional* object. The next step is to somehow visualize moving the cube one unit in the direction of the 4th dimension, and producing the hypercube, also called tesseract. In a similar way one can arrive at the *hypersphere*, a 4-D sphere. But mathematics does not stop with the 4th dimension, but considers the *nth-dimension*. Startling mathematical patterns appear when data referring to vertices, edges, and faces of different dimensional objects is collected and organized.

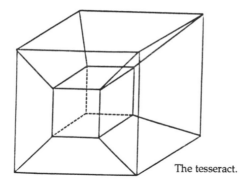

The tesseract.

The possible existence of a 4th-dimension has intrigued many people. Artists and mathematicians have tried to image and draw how certain objects would appear in the 4th dimension. The tesseract and the hypercube are 4th-dimensional representations of a cube. A cube drawn on paper is sketched in perspective to imply its 3-dimensional characteristic. Thus a tesseract drawn on paper is a perspective of a perspective.

Computers & Dimensions

Humans, being 3-D creatures, can easily visualize and understand the first three dimensions. Although dimensions exist mathematically beyond the third, it is difficult to accept something we cannot see or imagine. Computers are beginning to be used to help in the visualization of higher dimensions. For example, Thomas Bancroft (a mathematician) and Charles Strauss (a computer scientist) at Brown University have used a computer to generate motion pictures of a hypercube moving in and out of a 3-D space, thereby capturing various images at different angles of the hypercube in a 3-D world. It is analogous to a cube (i.e. 3-D object) passing through a plane (2-D world) at different angles and recording the cross-section impressions it makes on the plane. A collection of these impressions will help give a better picture to a 2-D creature of a 3-D object.

This illustration shows different impressions a sphere would leave as it passes or intersects a planes, i.e. the 2nd dimension. It is analogous to a hypercube passing through space, i.e. the 3rd dimension.

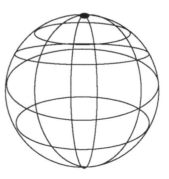

We now have 2-D holograms picturing 3-D objects. Holograms are now being commercially used in advertising and graphics. Perhaps in the future, 3-D holograms will be developed and used to picture 4-D objects.

Have you considered that even your best friend could be a 4th-dimensional creature who appears to you as 3-D creature?

$T_{opology}$ studies the properties of an object that remain unchanged when the object is distorted (stretched or

The "Double" Moebius Strip

shrunk). Unlike Euclidean geometry, topology does not deal with size, shape, or rigid objects. Basically, it studies elastic objects, which is why it has come to be referred to as *rubber sheet geometry*. The Moebius strip was created in the 17th century by the German mathematician Augustus Moebius, and is one of the objects studied in topology. By taking a strip of paper, giving it a half twist and gluing its ends together, a Moebius strip is formed. It is fascinating because it has only one side. A pencil can be used to trace its entire surface without lifting the pencil.

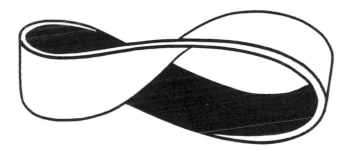

Now let's consider the *double Moebius strip*. It is formed by taking two strips of paper, simultaneously giving them a half twist and gluing the ends together. The strips appear to be two nestled Moebius strips. But are they?

Make the illustrated model and test it. Run your finger between the two strips to see if they seem nestled. Take a pencil and begin tracing along one of them, until you arrive back at your starting place. What happened?

What happens if you try to unnestle them?

Paradoxical Curve – space-filling curve

A curve is usually considered one-dimensional, and composed of points which are considered to have zero or no dimension. With these notions it somehow seems contradictory that a curve can fill a given space. Euclidean curves are planar, flat. Mathematicians of the Euclidean era had not yet considered that curves could generate themselves in the manner below.

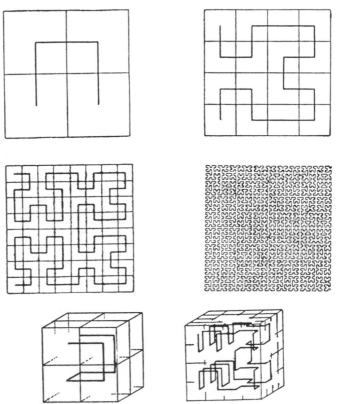

The example above shows the stages of the *space-filling curve* which will envelop the space of an entire cube by continually generating itself in the special way illustrated.

Often called an ancient computer, the abacus is one of the oldest accounting devices

The Abacus

known. This ancient calculator was and is used in China and other Asian countries to add, subtract, multiply, divide and calculate square and cube roots. There are many types of abaci. For example, the Arabic abacus had ten balls on each wire with no center bar. History shows that even the ancient Greeks and Romans used abaci.

The Chinese abacus consists of thirteen columns of beads divided by a crossbar. Each column has five beads below the crossbar and two beads above it. Each bead above the crossbar is equal to five beads below the cross bar in its column. For example, a bead above the crossbar in the ten's column is worth 5x10 or 50.

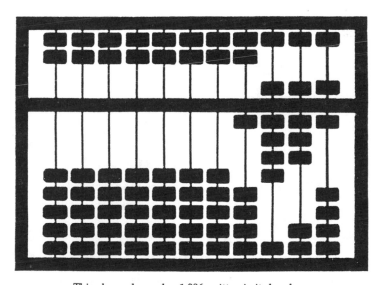

This abacus has value 1,986 written in its beads.

Mathematics & Weaving

How can mathematical objects be present in woven fabrics?

Does a weaver consciously analyze the design from a mathematical point of view.

Studying these pieces of fabric one finds many mathematical concepts present in them:

- lines of symmetry
- tessellations
- geometric forms
- proportional objects
- reflected patterns

Shoshone Indian design

Ojibwa Indian design

design from the Congo

Potawatomi Indian design

Can you find the mathematical ideas listed above in these fabrics' designs? Can you find other mathematical ideas present?

In February of 1984, a team of mathematicians successfully programmed their Cray computer (able

Mersenne's Number

to sample whole clusters of numbers simultaneously) to solve a three century old puzzle. The 17th century 69 digit number proposed prime by French mathematician Marin Mersenne was factored after 32 hours and 12 minutes of computer time by discovering the three factors listed below. This feat now has cryptographers worried, since many cryptographic systems use multidigit numbers, which are difficult to factor, in order to encode secrets and keep them secure.

MERSENNE'S NUMBER

13268610439897205317760857550609056142935393589033352580 2891469459697
factors:
178230287214063289511   *and*   61676882198695257501367   *and* 120703961782498930399969681

Factoring a number involves breaking it down into a product of smaller prime numbers. The task is simple with small numbers, and can be done by trying all primes numbers smaller than the number as divisors. Large numbers require other mathematical methods. The number of computations needed to factor a number grows exponentially as the numbers grow. Even a computer performing a billion operations per second would take several thousand years to factor a 60-digit number by the above method.

In 1985-86, Robert Silverman (of Mitre Corp. Bedford, MA) and Peter Montgomery (of System Development Corp. Santa Monica, CA) developed a method of factoring using microcomputers, rather than special computers developed expressly for factoring or the costly Cray computer. Their method is fast and very inexpensive. One of their recent accomplishments was factoring an 81-digit number using eight microcomputers which ran for 150 hours.

Tangram Puzzle

Using the seven pieces of a tangram, determine how the figures below are formed.

$\mathbf{T}$his illustration shows how a set of points of a semicircle, which is finite in length, is matched (put into a one-to-one correspondence)

Infinite vs Finite

with a set of points of a line, which is infinite in length. This semicircle has a perimeter of 5π. The line tangent to the semicircle is infinite in length. A ray with endpoint P (center of the semicircle) is drawn so that it intersects the line and the semicircle. The points of intersection of the semicircle and the line are thus matched up by the ray in a one-to-one correspondence. As the ray moves along the semicircle and approaches ray PQ, it intersects the line further and further out.

What would happen when the ray becomes ray PQ? [1]

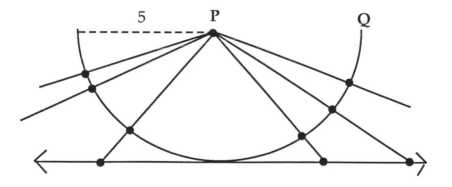

---

[1]The ray would be parallel to the line.

Triangular, Square & Pentagonal Numbers

There are many different labels given to numbers. Some of their names were derived because of the shape of geometric objects they form. Below we see that *odd numbers* form triangular shapes and hence they are also called *triangular numbers. Perfect square numbers*, i.e. $1^2=1$, $2^2=4$, $3^2=9$, ..., form squares.

Each group of numbers has a pattern associated with it. Try to form other sequences of numbers that are linked to geometric objects, and determine their particular pattern.

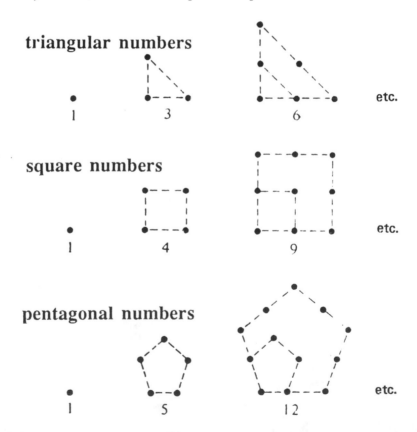

triangular numbers

1   3   6   etc.

square numbers

1   4   9   etc.

pentagonal numbers

1   5   12   etc.

| In 200 B.C., Eratosthenes devised this ingenious method for measuring the distance around the Earth. | *Eratosthenes measures the Earth* |

To measure the circumference of the Earth, Eratosthenes used his knowledge of geometry, particularly the theorem:

> *Parallel lines cut by a transversal*
> *form congruent alternate interior angles.*

He was aware that at noon during the summer solstice in the city of Syene (in Egypt) a vertical rod did not cast a shadow, while in Alexandria (5000 stadia ≈ 500 miles away) the vertical rod cast a shadow which formed a 7º 12' angle. With this information he was able to calculate the circumference of the Earth to within 2% of its actual value.

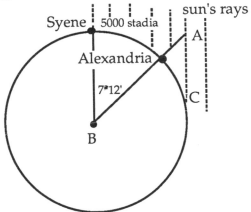

procedure:
Since light rays travel parallel to each other, <CAB and <B in the above diagram are congruent alternate interior angles.

Thus, the distance between Syene and Alexandria is a fraction of the distance around the Earth. This fraction is 7º 12'/ 360º = 1/50. Therefore, the distance around the Earth is (500 miles) x 50 = 25,000 miles.

Projective Geometry & Linear Programming

Using techniques of projective geometry and solving systems of equations, **Narendra Karmarkar**, while a Bell Laboratory mathematician, discovered a method to drastically cut down the time needed to solve very cumbersome linear programming problems, such as allocating time on communication satellites, scheduling flight crews, and routing millions of telephone calls over long distances.

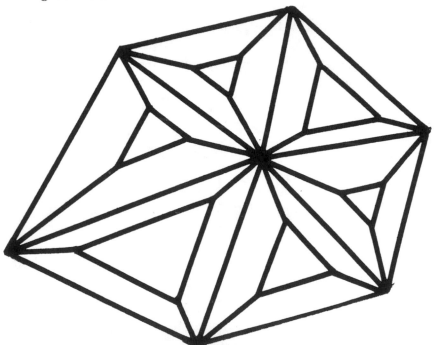

An artist's rendition of a geometric solid and its many facets.

Until recently, the *simplex method*, developed by mathematician George B. Danzig in 1947, had been used. It requires large amounts of computer time and is not feasible for enormous problems. Mathematicians visualize such problems as complex geometric solids with millions or billions of facets. Each corner of each facet represents a possible solution. The task of the algorithm[1] is to find the best solution without having to calculate every one. The simplex method by Danzig runs along the edges of the solid, checking one corner after another but always heading for the best solution. In most problems it manages to get there efficiently enough, as long as the number of variables (unknowns) is no more than 15,000 to 20,000. The *Karmarker algorithm* takes a short-cut by going through the middle of the solid. After selecting an arbitrary interior point, the algorithm warps the entire structure, i.e. reshaping the problem, in a way designed to bring the chosen point exactly into the center. The next step is to find a new point in the direction of the best solution and to warp the structure again, and bring the new point into the center. Unless the warping is done, the direction that appears to give the best improvement each time will be an illusion. The repeated transformations, based on concepts of projective geometry, lead rapidly to the best solution.

---

[1]An algorithm is a procedure of computation for arriving at a solution. For example, the process and steps of long division is an algorithm. In long division we mentally take short-cuts in the division process—if 29 is to be divided into 658, one thinks of 29 as being close to 30 and figures how many 30's are in 65 rather than trying to figure out how many 29's in 658 all at once. The *Karmarker algorithm* also has specific short-cuts built into it, as its transformation/warping process.

The spider & the fly problem

Henry Ernest Dudeney was a renowned 19th century English puzzle creator. Most of today's puzzle books have many of his gems, but very often they are not credited to him. In the 1890's he and Sam Loyd, the famous American puzzlist, collaborated on a series of puzzle articles.

Dudeney's first book, *The Canterbury Puzzles,* was published in 1907. Five additional books were later published and they remain a treasury of mathematical teasers.

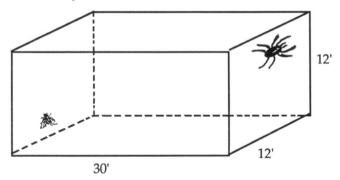

The spider and the fly, which first appeared in an English newspaper in 1903, is one of his best known puzzles.

In a rectangular room, 30'x12'x12', a spider is at the middle of an end wall, one foot from the ceiling.

The fly is at the middle of the opposite end wall, one foot above the floor. The fly is so frightened it can't move.

What is the shortest distance the spider must crawl in order to capture the fly? (Hint: it is less than 42')

See appendix for the solution to *the spider & the fly problem.*

Whathat kind of mathematical concepts would be connected to soap bubbles? The shapes that soap film forms are governed by surface tension.

Mathematics & soap bubbles

The surface tension diminishes the surface area as much as possible. Consequently, each soap bubble encloses a certain amount of air in such a way that the surface area for the given amount of air is minimized. This explains why a single soap bubble becomes a sphere, while a cluster of bubbles, as in foam, has a different formation. In foam, the edges of the soap bubbles meet at angles of 120º, called *triple junctions*.   A triple junction is essentially the point where three line segments meet, and the angles at the intersection are 120º each.   Many other natural phenomena (a few examples are the scales of a fish, the interior of a banana, formation of kernels of corn, the plates on a turtle's shell) conform to a triple junction, which is an equilibrium point of nature.

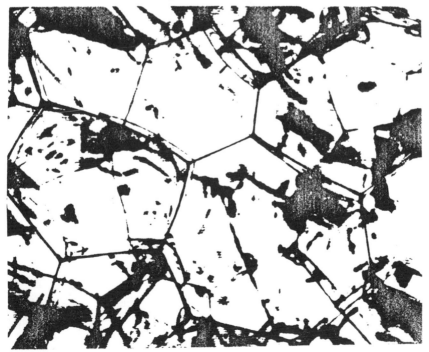

The coin paradox

The top coin has been moved halfway around the coin below it ending up in the same position in which it originally started. Since it traveled half of its circumference, one would have expected it to end upside down. Take two coins and study the movement. Can you explain why this did not happen?

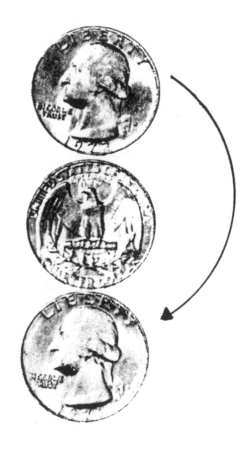

Hexaminoes

An hexamino is a flat object composed of six square units. Starting with a cube of volume 1 cubic unit, cut along seven of its edges so it falls apart and lies flat. The resulting figure is a hexamino. Depending along which edges the cube is cut, different shaped hexaminoes result. Some hexaminoes are pictured below.

How many different hexaminoes are there?

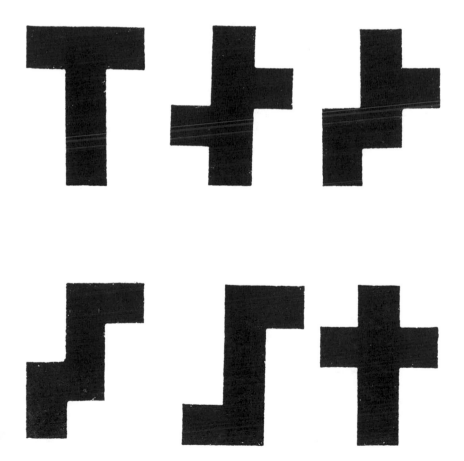

The Fibonacci sequence & Nature

The occurrence of the Fibonacci sequence in nature is so frequent that one is convinced it cannot be accidental.

a) Consider the list of the following *flowers with a Fibonacci number of petals:*   trillium, wildrose, bloodroot, cosmos, buttercup, columbine, lily blossom, iris.

b) Consider these *flowers with a Fibonacci number of petal-like parts:* aster, cosmos, daisy, gaillardia.

these Fibonacci numbers are frequently associated with the petals of:

> 3.................lilies and irises
> 5.................columbines, buttercups, and larkspur
> 8.................delphiniums
> 13...............corn marigolds
> 21...............asters
> 34, 55, 84...daisies

BLOODROOT

TRILLIUM

COSMOS

WILD ROSE

c) Fibonacci numbers are also found in the *arrangement of leaves, twigs,* and *stems.* For example, select a leaf on a stem and assign it the number 0, then count the number of leaves (assuming none have been broken-off) until you reach one directly in line with the 0-leaf. The total number of leaves will most likely be a Fibonacci number, and the number of turns in the spiral before reaching the leaf directly above should also be a Fibonacci number. The ratio of leaves to turns in the spiral is called a phyllotactic (from the Greek word meaning *leaf-arrangement*) ratio. Most phyllotactic ratios happen to be Fibonacci ratios.

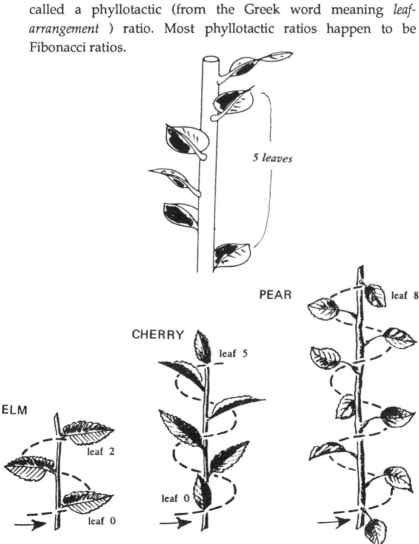

5 leaves

ELM

PEAR

CHERRY

d) Fibonacci numbers have sometimes been called the *pine cone numbers* because consecutive Fibonacci numbers have a tendency to appear as left and right sided spirals of a pine cone. This is also true for a sunflower seedhead. In addition, you may find some that are consecutive Lucas numbers.[1]

8 spirals to the right and 13 spirals to the left

sunflower seedhead

---

[1] Lucas numbers form a Fibonacci-like sequence which starts with the numbers 1 and 3; then consecutive numbers are obtained by adding the previous two numbers. Thus, the Lucas sequence is 1,3,4,7,11, . . .It is named after Edouard Lucas, the 19th century mathematician who gave the Fibonacci sequence its name and who studied recurrent sequences. Another way in which the Lucas sequence is related to the Fibonacci sequence is the following:

$$0, \ 1, \ 1, \ 2, \ 3, \ 5, \ 8, \ 13, \ \dots$$
$$1, \ 3, \ 4, \ 7, \ 11, \ 18, \ \dots$$

e) The *pineapple* is another plant to check for Fibonacci numbers. For the pineapple, count the number of spirals formed by the hexagonal shaped scales on the pineapple.

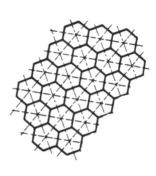

Fibonacci sequence & the golden ratio

The sequence of consecutive Fibonacci ratios,

$$\frac{1}{1}, \ \frac{2}{1}, \ \frac{3}{2}, \ \frac{5}{3}, \ \frac{8}{5}, \ \cdots, \ \frac{F_{n+1}}{F_n}, \ \cdots$$

$$1, \ 2, \ 1.5, \ 1.6, \ 1.625, \ 1.6153, \ 1.619, \ \ldots$$

alternates above and below the value of the golden ratio, φ. The limit of this sequence is φ. This connection implies that wherever (particularly in natural phenomena) the golden ratio, the golden rectangle, or the equiangular spiral appear the Fibonacci sequence is present and vice versa.

The monkey and the coconuts

Three shipwrecked sailors and a monkey found themselves on an island where the only food was coconuts. They gathered coconuts all day, and decided to go to sleep and divide them up the next day. During the night one of the sailors woke up and decided to take his share of the coconuts rather than wait until morning. He divided the coconuts into three piles, but there was one coconut left over which he gave to their monkey. He hid his pile and went back to sleep. Later, another sailor got up and did the same thing as the first sailor, giving the leftover coconut to the monkey. And later on the third sailor woke up and proceeded to divide the coconuts as the other two sailors had done, also giving the left over coconut to the monkey. In the morning, when the three sailors got up, they divided the pile of coconuts into 3 shares with one left over for the monkey.

What is the least number of coconuts the sailors collected?

Try the same problem with four sailors, and then with five sailors.

The equations used to solve these problems are called Diophantine, after the Greek mathematician Diophantus who first used these types of equations for certain types of problem solving.

See the appendix for the solution to *the monkey & the coconuts problem.*

Spider & Spirals

Four spiders start crawling from four corners of a 6x6 meter square. Each spider is crawling toward the one on its right, moving toward the center at a constant rate of 1 centimeter per second. Thus, the spiders are always located at four corners of a square.

How many minutes will it take to meet at the center?

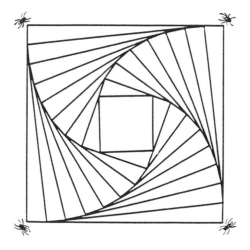

The curves formed by the spiders' paths are *equiangular spirals.*[1]

Try reworking the problem with other shaped regular polygons.

See appendix for the solution to *the spider & spiral.*

---

[1]For additional information on the equiangular spiral, see the section on **The Golden Rectangle.**

APPENDIX

solutions • answers • explanations
• results • solutions • answers •
explanations • results • solutions
• answers • explanations • results
• solutions • answers • explanations • results • solutions • answers • explanations • results •
solutions • answers • explanations
• results • solutions • answers •
explanations • results • solutions
• answers • explanations • results
• solutions • answers • explanations • results • solutions • answers • explanations • results •
solutions • answers • explanations
• results • solutions • answers •

SOLUTIONS

page 9—*triangle to a square*

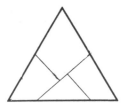

page 17—*the wheat and the chessboard*

$1 + (2) + (2)^2 + (2)^3 + (2)^4 + ... + (2)^{63}$

$1 + 2 + 4 + 8 + 16 + ...$

page 35—*the T problem*

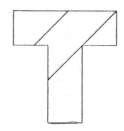

page 37—
He decided to shift each occupant to a room that had a number twice as large, so the guest with room #1 went to #2, #2 went to #4, #3 went to #6, etc. This opened up all the odd numbered rooms for the infinite bus load of guests.

page 47—*a Sam Loyd puzzle*
Starting at the center move the appropriate squares in the following direction patterns: SW, SW, NE, NE, NE, SW, SW, SW, NW.

page 51—*Fibonacci trick*
If a and b represent the first two terms, then the following terms generated would be: a+b; a+2b; 2a+3b; 3a+5b; 5a+8b; 8a+13b; 13a+21b; 21a+34b. Now taking the sum of the first ten terms gives 55a +88b, which is 11 times the seventh term, 5a+8b.

page 56—*ten histroical dates:*
1879-birth of Einstein; 1066-Battle of Hastings; 476-Fall of Rome; 1215-Magna Carta; 1455-Gutenberg Bible; 563-Birth of Buddha; 1770-birth of Beethoven; 1969-walk on the moon; 1948-Ghandi assasinated; 1776-U.S.Declaration of Independence.

SOLUTIONS

page 59 —*problem #8 from Pillow Problems*
Lewis Carroll's explanation—
> m=# of men
> k=# of shillings of the last man (the poorest man)

After once around, each man is one shilling poorer , and the moving pile of shillings contains m shillings. After k times around each is k shillings poorer, with the last man having no shillings and the heap containing mk shillings. The process ends when the last man must pass on the pile, which then contains (mk+m-1) shillings. The next to the last man now has nothing, and the first man has (m-2) shillings.

The first and the last man are the only two neighbors whose shillings can be in a ratio of 4 to 1. Thus either
> mk+m-1=4(m-2)   or else
> 4(mk+m-1)=m-2.

The first equation gives mk=3m-7, i.e. k=3-(7/m), which gives no integral values other than m=7, and k=2.
The second equation gives 4mk=2-3m, which gives no positive integral values.
Thus the answer is 7 men and 2 shillings.

page 69
Proof for *the amazing track:*
The area of the race track is — $\pi R^2 - \pi r^2$.
This is the difference of the large circle's area minus the small circle's area.

From the diagram on p. 69 we see that the chord's length is $2\sqrt{(R^2 - r^2)}$.

Thus a circle with this diameter would have area $\pi (R^2 - r^2)$, which simplifies to $\pi R^2 - \pi r^2$.

page 70 —*Persian Horses*
There are two horses horizontally placed belly to belly, and two horses vertically placed belly to belly.

page 71—*Sam Loyd's horses*

SOLUTIONS

page 116—*Achilles and the tortoise*
Achilles would reach the tortoise at one-thousand one hundred and eleven and one-ninth meters. If the race track is shorter than this, the tortoise would win. If it were exactly this size, it would be a tie. Otherwise, Achilles will pass the tortoise.

page 123—*Diophantus' riddle*
n represents the number of years Diophantus lived.
(1/6)n+ (1/12)n + (1/7)n +5 + (1/2)n +4 =n
simplifying: (3/28)n = 9
$$n = 84 \text{ years old}$$

page 136—*the checkerboard problem*
It is not possible to cover the altered checkerboard with dominoes. A domino must occupy a red and black square. Since both corners removed were the same color, there will not be a compatible number of red and black squares left.

page 140—*The proof of 1=2?*
Division by zero occurs at step 6. The number zero is camouflaged by being expressed as b-a. b-a equals zero because a=b in the premise.

page 147—*the unexpected exam paradox*
The test cannot be on Friday because it is the last day it could be given and you could deduce this on Thursday, if you had not had it yet. The condition was that you would not know which day until the morning of the test. So if it were not Friday, that would make Thursday the last possible day. But it cannot be on Thursday because by Wednesday you would know there are Thursday and Friday left. Since Friday was out, on Wednesday you would know ahead of time it would be Thursday, which you are not supposed to know ahead of time. Now this leaves Wednesday as the last possible day, but Wednesday is out because if you did not have it by Tuesday, you would know by Tuesday you would have it on Wednesday. Continuing with this reasoning, each day of the week is eliminated.

page 159—*farmer, wolf, goat & cabbage*
He first takes the goat across. He then returns and picks up the wolf. He leaves the wolf off, and takes the goat back. He then leaves the goat at the starting place, and takes the cabbage over to where the wolf is. He then returns, and picks up the goat, and goes where the wolf and the cabbage are waiting.

page 163—*nine coin puzzle*

SOLUTIONS

page 175—*wood, water grain problem*

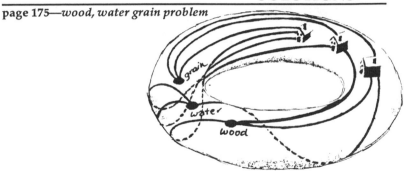

The *wood, water and grain problem* has no solution so long as the paths must remain on the (Euclidean) plane. But if the houses are situated on a *torus* or doughnut surface (as illustrated) the solution is simple.

page 181—*counterfeit coin puzzle*
One weighing!
Place on the scale one coin from the first stack, two from the second, three from the third, and so forth. You know how much the coins on the scale should weigh if none of them were counterfeit. Therefore to determine which stack is counterfeit, look at the weight and figure out how much heavier it weighs. This number will correspond to the stack from which that group of coins came. For example, if it's four grams heavier, the forth stack is counterfeit because you placed four of its coins on the scale.

page 190—*three men facing the wall*
The man furthest from the wall either sees 2 tan hats or a black and tan hat. If he saw 2 blacks hats, he would have known he had to have a tan hat. The middle man sees a tan hat because if he saw a black hat, he would known he must be wearing a tan hat from the first response. Therefore the man facing the wall concludes he can only be wearing the tan hat the middle man sees.

page 218—*the spider and the fly*

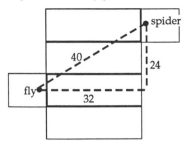

SOLUTIONS

page 227—*the monkey and the coconuts*

seventy-nine coconuts. Let n represent the amount of original coconuts.

# for monkey	# each sailor hid for himself	# left in pile
1	(n-1)/3	(2n-2)/3
1	[(2n-5)/3]+3=(2n-5)/9	2(2n-5)/9=(4n-10)/9
1	[(4n-19)/9]+3=(4n-19)/27	2(4n-19)/27=(8n-38)/27

	# each sailor took in morning	
1	[(8n-65)/27]+3=(8n-65)/81	0

Recall n=total original # of coconuts.
(8n-65)/81=f, the number of coconuts each sailor received when the coconuts were divided in the morning. Let f take one the various successive values of the counting numbers, starting with 1. The smallest counting number that gives a whole number value for n is f=7. With this value, n has value 79.

page 228—*spider and spirals*
Notice as the spiders move the size of the square they form shrinks, but it always remains a square. Each spider's path is perpendicular to the spider's path on its right. A spider will reach the spider on its right in the same time it would take if the spider on the right had not moved. Each spider will have traveled 6 meters, which is 600 centimeters. 600 centimeters will have taken 600 seconds or 10 minutes.

INDEX

INDEX

INDEX

————————————————About The Author ————————————

Mathematics teacher and consultant Theoni Pappas received her
B.A. from the University of California at Berkeley in 1966 and
her M.A. from Stanford University in 1967. She is committed to
demystifying mathematics and to helping eliminate the elitism
and fear that often are associated with it.

In addition to *The Mathematics Calendar*, her other innovative
creations include *The Math-T-Shirt, The Children's Mathematics
Calendar, The Mathematics Engagement Calendar,* and *What Do You
See?*—an optical illusion slide show with text. Ms Pappas is also
the author of the following books: *Mathematics Appreciation,* and
Greek Cooking for Everyone.